I0037203

Materials Joining and Manufacturing Processes
MJMP 2025

International Symposium on Recent Advances in
Materials Joining and Manufacturing Processes,
27th and 28th February 2025, held by the Dept. of Mechanical
Engineering, Alva's Institute of Engineering and Technology,
Moodbidri, Karnataka, India

Editors
**Satyanarayan[1], Kazuyuki Hokamoto[2],
Suresh P.S.[1] and Kumar Swamy M.C.[1]**

[1]Department of Mechanical Engineering, Alva's Institute of Engineering and
Technology, Karnataka, Moodbidri, Mangalore 574225 (affiliated to Visvesvaraya
Technological University, Belagavi, Karnataka, India)

[2]Institute of Industrial Nanomaterials, Kumamoto University, Japan 8608555

Peer review statement

All papers published in this volume of "Materials Research Proceedings" have
been peer reviewed. The process of peer review was initiated and overseen by the
above proceedings editors. All reviews were conducted by expert referees in
accordance to Materials Research Forum LLC high standards.

Copyright © 2025 by authors

[cc] BY Content from this work may be used under the terms of the Creative Commons Attribution 3.0 license. Any further distribution of this work must maintain attribution to the author(s) and the title of the work, journal citation and DOI.

Published under License by **Materials Research Forum LLC**
Millersville, PA 17551, USA

Published as part of the proceedings series
Materials Research Proceedings
Volume 55 (2025)

ISSN 2474-3941 (Print)
ISSN 2474-395X (Online)

ISBN 978-1-64490-361-2 (eBook)
ISBN 978-1-64490-360-5 (print)

This book contains information obtained from authentic and highly regarded sources. Reasonable efforts have been made to publish reliable data and information, but the author and publisher cannot assume responsibility for the validity of all materials or the consequences of their use. The authors and publishers have attempted to trace the copyright holders of all material reproduced in this publication and apologize to copyright holders if permission to publish in this form has not been obtained. If any copyright material has not been acknowledged please write and let us know so we may rectify in any future reprint.

Distributed worldwide by

Materials Research Forum LLC
105 Springdale Lane
Millersville, PA 17551
USA
https://mrforum.com

Manufactured in the United State of America
10 9 8 7 6 5 4 3 2 1

Table of Contents

Preface

Committees

Study on the milling of cobalt-chromium (Co-Cr) alloy produced by wire arc additive manufacturing (WAAM)
Gautama Hebbar A., Srinivasa Pai P., Dinesh Singh Thakur, Vijeesh Vijayan,
Bhaskara P. Achar, Vikas Marakini 1

Effect of heat treatment on mechanical and wear properties of aluminum tin (Pb free) bearing material
Bhat Jayarama, Satyanarayan, B.H. Vadavadagi, P.S. Shivakumar Gouda,
Algur Veerabhadrappa... 7

Development of internal cooling system in hard turning of annealed HCWCI using carbide tools
A.M. Ravi, S.M. Murigendrappa.. 12

Investigation of mechanical properties of natural material 3D printed specimens
Vinay Swamy, Gurupadayya, Naveen Kumar, Vishalagoud S. Patil, Gurushant B. Vaggar,
Dodda Hanamesha.. 19

Effect of inclination angle in explosive welding of magnesium-zinc sheets: A numerical approach
Samuel Debbarma, Subrata K. Ghosh, S. Saravanan, Prabhat Kumar.. 27

Evaluating equipment-dependent forging behavior of aluminum alloy
Akash Mahato, Saurabh Shrivastava, Sujit Goswami, Suman Kumar Pandey,
Rahul Ramesh Kulkarni 34

Modification of dye/fluorescent penetrant testing in accordance with Industry 4.0
Nitish Kumar, Banshidhara Mallik, Rahul Ramesh Kulkarni 40

Mechanical performance of glass fiber epoxy laminates with embedded circular and square cutouts
Aravind MUDDEBIHAL, P.S. Shivakumar Gouda, Vinayak S. UPPIN, I. Sridhar 45

Optimization of process parameters in explosive welding using machine learning
Kusammanavar Basavaraj, Satyanarayan, Anand Kulkarni... 51

An investigation of thermogravimetric analysis and thermal conductivity of glass fibre epoxy resin composites modified with silicon carbide, manganese, and copper nanoparticles (NPs)
Gurushanth B. Vaggar, S.C. Kamate2,b, Kiran C.H., Deepak Kothari,
S.L. Nadaf, Vishalagoud Patil 57

A review on effect of filler materials on thermal properties of hybrid polymer matrix composites
Gurushanth B. Vaggar, S.L. Nadaf, Narayan V., Dharshith, Praveen V. 64

Review on additive manufacturing at the forefront: Exploring recent developments and industry applications
Gurushanth B. Vaggar, Elvin Chris Dsouza, Shyam, Ananthesh D. Kamath............................. 72

Implosive reactive armour to enhance explosive welding on hard surfaces in modern warfare
A. Gyanesh Kumar Rao, Satyanarayan, Suresh Ganesh Kulkarni 79

Exploring advanced micro-forming methods: Ultrasonic vibration (UV), pulsed current, and laser-assisted approaches
Mihiretu Ganta, Mahaboob Patel, Tewodros Dawit 85

Influence of tool rotation speed on mechanical and microstructural characteristics of friction stir weld 7075 aluminum alloy reinforced with graphene nano platelets
Rahul Biradar, Sachinkumar Patil 95

Investigation on mechanical and microstructural properties of friction stir weld AA8xxx series alloy: A review
Maheshwarayya K.C., Sachinkumar PATIL, Mahesh L. 102

Structural evaluation of microwave butt welded MONEL 400 sheets through finite element analysis
Gajanan M. Naik, Shanthala K., Sadashiv Bellubbi, Chirag L., Devendra Gowda,
Suresh Poyil Subramanyam 110

Optimization of mechanical properties in 3D-printed PLA parts with honeycomb and cubic infill patterns using taguchi method
Aveen K.P., Shivaramu H.T., Praveen K.C., Sridhar D.R., Thoran 118

Joining of aluminum tube to PVC pipes through electromagnetic force
Shanthala K., Gajanan M. Naik, Akshit Kochhar, Ramesh S., Sadashiv Bellubbi 130

Lead-free solders for high-temperature applications
SATYANARAYAN, K. NARAYAN PRABHU 136

Thermal spray coatings in industrial boiler environments: A review
Jayson Anil Pinto, K. Narayan Prabhu 141

Characterization and thermal performance of Sn-Bi alloy used as a thermal interface material
Kumar Swamy M.C., Satyanarayan 148

Mechanical performance of 3D-printed PBAT composites reinforced with sawdust
Maruthi Prashanth B.H., P.S. Shivakumar Gouda, Sandeepkumar Gowda,
Asif Iqbal Mulla, Girish Ariga, Suresh P.S., Sandeep Kyatanavar, Srinivasa C.S. 156

Keyword Index
About the Editors

Preface

Proceedings of the International Symposium (MJMP 2025) on "Recent Advances in Materials Joining and Manufacturing Processes"

The Department of Mechanical Engineering of Alva's Institute of Engineering and Technology (AIET), Moodbidri, proudly hosted the International Symposium on "Recent Advances in Materials Joining and Manufacturing Processes (MJMP 2025)" on 27th and 28th February 2025 at the AIET campus, India. This landmark event brought together distinguished scientists, researchers, technologists, and academicians from across India and abroad to exchange insights on evolving trends in materials joining (such as welding, brazing and soldering), additive manufacturing, composite manufacturing, and simulation techniques.

The symposium was organized in collaboration with the Institute of Industrial Nanomaterials (IINa), Kumamoto University, Japan, one of Japan's oldest and most prestigious institutions, and BETA CAE Systems India Pvt. Ltd., Bengaluru, a leader in simulation solutions. The event witnessed participation from India and across the globe, reflecting the national and international footprint and academic relevance of the symposium. Symposium received a total of 23 high−quality research papers for presentation and publication, ensuring a rich blend of academic and industrial perspectives.

A key highlight of the symposium was the diverse panel of 20+ eminent speakers, including experts from Japan, DRDO, CPRI, HAL, Airbus, MRPL and renowned Indian institutions such as NITK, DIAT, NIFFT and Annamalai University. Research scholars from Poland, Ethiopia and premier institutions from India presented their latest results in connection to advanced materials joining, additive manufacturing, interface materials, fastening technology, and coating technology in terms of poster presentation sessions. Even AIET's faculties contributed significantly, underlining the institute's commitment to global research excellence. Organising chairs extended sincere gratitude to all the keynote speakers, contributors, reviewers, advisory committee members, sponsors and organising committee members for their valuable inputs and ongoing dedication to making this symposium a success. It is confident that the deliberations and knowledge shared during the event were inspired and will inspire research and innovation in the field of materials joining and manufacturing.

Committees

PATRONS

Chief Patron : **Dr. M. Mohan Alva,** Chairman, AEF
Patron : **Mr. Vivek Alva**, Managing Trustee, AEF
Patron : **Dr. Peter Fernandes**, Principal, AIET

ORGANIZING CHAIRS

Dr. Satyanarayan, Professor and Head, Dept. of ME, AIET, India
Dr. Kazuyuki Hokamoto, Professor, IINa, Kumamoto University, Japan

COORDINATORS

Dr. Suresh P S, Dept. of ME, AIET, India
Dr. Kumar Swamy M C, Dept. of ME, AIET, India

ADVISORY BOARD

Dr. Richard Pinto, Professor Emeritus, Dean R & D, AIET.
Dr. Divakara Shetty S, Dean - Academics, AIET.
Dr. Dattathreya, Dean - Planning & Development, AIET.
Prof. Durgaprasad Baliga, Dean - Student Affairs, AIET.
Dr. K V Suresh, Head, Agriculture Engineering Department, AIET.
Mr. Stavros Kleidarias, CEO, BETA CAE Systems India Pvt. Ltd., Bengaluru (B'lore).
Mr. Lokesh S Bommani, Head - Sales, India & Southeast Asia Region, BETA CAE Systems India Pvt. Ltd., B'lore.
Mr. Nagananda Upadhya, Head - Technical Support, BETA CAE Systems India Pvt. Ltd., Bengaluru.

KEYNOTE SPEAKERS

Dr. K.N Prabhu, NITK, Surathkal
Dr. Shuichi Torii, Kumamoto University, Japan
Dr. K Raghukandan, Annamalai University
Mr. V V Kamath, Fronius India Pvt. Ltd.
Dr. M G Anandakumar, CPRI, Bengaluru
Dr. B B Sherpa, Kumamoto University, Japan
Dr. Richard Pinto, AIET, Moodbidri
Dr. Somasundaram Saravanan, Annamalai University
Dr. Rahul Ramesh Kulkarni, NIAMT, Ranchi
Dr. Satyanarayan, AIET, Moodbidri

Dr. K. Hokamoto, Kumamoto University, Japan
Dr. Suresh G Kulkarni, DIAT, Pune
Dr. Pal Dinesh Kumar, TBRL, DRDO, Chandigarh
Dr. R Tomoshige, Sojo University, Japan
Mr. George Suraj Dsa, Expert, Airbus, Bengaluru
Dr. Shashikantha Karinka, AIET, Moodbidri
Dr. B Kusammanavar, RYMEC, Ballari
Mr. Rahul Raje, Fronius India Pvt. Ltd.
Mr. Jayson Anil Pinto, MRPL, Mangaluru
Dr. Vijesh V, NMAMIT, Nitte

TECHNICAL ADVISORY COMMITTEE

Dr. Gao Xin, Beijing Institute, China
Dr. Shigeru Tanaka, Kumamoto University, Japan
Dr. Shivaprasad K, Durham University, UK
Dr. S Janakiraman, IIT, Indore
Dr. Rajasekharan, Jazan University, Saudi Arabia
Dr. Raju, SJCE Mangalore
Dr. P S Shivakumar Gouda, SDMCET, Dharwad
Dr. Girish Kumar, SDMIT, Ujire

Dr. Vishwanath H M, MIT, Manipal
Dr. Sadashiv Bellubbi, JCER, Belagavi
Prof. Vani Adiga, YIT, Moodbidri
Dr. Ramakrishna D P, SCE, Mangalore
Dr. A M Ravikumar, GPT, Davanagere
Dr. Veerabhadrappa Algur, RYMEC, Ballari
Dr. Prakrathi S, MSRIT, Bangalore
Prof. Jayaram Bhat, SDMCET, Dharwad

ORGANIZING COMMITTEE

Prof. Veerendra Kumar, AIET, Moodbidri
Dr. Gurushanth B Vaggar, AIET, Moodbidri
Prof. Sharathchandra Prabhu, AIET, Moodbidri
Dr. Pramod V B, AIET, Moodbidri
Prof. Kiran C H, AIET, Moodbidri
Prof. Hemanth Suvarna, AIET, Moodbidri

Prof. Praveen K C, AIET, Moodbidri
Prof. Ganesh M R, AIET, Moodbidri
Prof. Deepak Kothari, AIET, Moodbidri
Prof. Srinivasa C S, AIET, Moodbidri
Mr. Nikhil Alva, AIET, Moodbidri
Mr. Avinash, AIET, Moodbidri

DISTINGUISHED REVIEWERS

Dr. Bir Bahadur Sherpa, Kumamoto University, Japan	Dr. P S Shivakumar Gouda, MITE, Moodbidri
Dr. Gangadhar K, JCER Belagavi	Dr. Kripa Suvarna, Cyient India Ltd, Bengaluru
Dr. Sadashiv Bellubbi, JCER, Belagavi	Dr. S. Prakrathi, RIT, Bengaluru
Dr. Mallikarjun Jalageri, JCER, Belagavi	Dr Mahantesh H M, RYMEC Ballari
Dr. Gurushanth B Vaggar, AIET, Moodbidri	Dr. Ramakrishna D P, SCE, Mangalore
Dr. Suresh P S, AIET, Moodbidri	Dr Abilash Desai, SDMCET, Dharwad
Dr. Vijayendra Kukanur, H.K.E Soci. SMVCE, Raichur	Dr. Pramod V B, AIET, Moodbidri
Dr. Gurushanth B Vaggar, AIET, Moodbidri	Prof. Vani R, YIT, Moodbidri

Materials Joining and Manufacturing Processes: MJMP 2025   Materials Research Forum LLC
Materials Research Proceedings 55 (2025) 1-6   https://doi.org/10.21741/9781644903612-1

Study on the milling of cobalt-chromium (Co-Cr) alloy produced by wire arc additive manufacturing (WAAM)

Gautama Hebbar A.[1,a] *, Srinivasa Pai P.[1,b] *, Dinesh Singh Thakur[2,c],
Vijeesh Vijayan[1,d], Bhaskara P. Achar[1,e], and Vikas Marakini[1,f]

[1]Department of Mechanical Engineering, Nitte (Deemed to be University), NMAM Institute of Technology (NMAMIT), Nitte, India, 574110

[2]Department of Mechanical Engineering, Defence Institute of Advanced Technology (DU), Pune, Maharashtra, India, 411025

[a]gouthamahebbar@nitte.edu.in, [b]srinivasapai@nitte.edu.in, [c]dinnu74@yahoo.com, [d]vijeeshv@nitte.edu.in, [e]bhaskarap@nitte.edu.in, [f]vikasbhat02@gmail.com

Keywords: Wire Arc Additive Manufacturing, Milling; Difficult-to-Cut Material, Surface Roughness, Tool Wear

Abstract: In recent years, Cobalt-chromium (Co-Cr) alloys have gained popularity especially in dental and orthopedic implants due to their exceptional properties related to bio compatibility. However, they have not been extensively studied. Wire arc additive manufacturing (WAAM) is a promising manufacturing technique, as it can produce parts with mechanical properties similar to wrought or cast materials. But the surface finish from WAAM produced components are questionable. Milling is widely used technique to improve the surface finish of WAAM produced component. Therefore, this work outlines the work carried out to suggest the best possible milling conditions and suitable tool inserts for milling WAAM'ed Co-Cr alloys. The WAAM'ed Co-Cr alloy is prepared by WAAM. The hardness of the alloy is found to be 783.5 HV. CVD coated inserts and PVD coated inserts have been used for machining. A cutting speed of 150 m/min, machining length of 100 mm and radial depth of cut (DoC) of 50 mm is initially set. Due to high hardness of the material the inserts could not machine in the set cutting conditions. Reducing the radial DoC and cutting speed are found to be safe to machine without causing tool damage. A combination of 50 m/min Cutting speed (V_c), 0.05 mm/z feed rate, 0.1 mm axial depth DoC, 5 mm radial depth of cut, and 50 mm machining length is recommended using PVD coated insert to produce better surface roughness (Ra = 0.0514 μm) and reduced tool insert flank wear (0.066 mm).

Introduction

Today's materials are being pushed to their limits by the natural desire for more efficient machinery and parts. New alloy systems' properties and technical suitability are continuously being studied in an effort to push these boundaries even further. In recent years, research has focused more on a new class of alloys known as cobalt-chromium (Co-Cr) alloys. Typically, cobalt, chromium, and additional elements like molybdenum, nickel, and iron make up Co-Cr alloys [1-2]. The stability, anti-corrosion, and biocompatibility of these alloys are superior to those of Ni-Cr and Ti alloys. These alloys' high specific strength, resistance to mechanical stress, and ability to withstand corrosive and abrasive loads make them popular in applications like dental and orthopedic implants [3,4]. The process of characterizing the potential properties of this alloy group is just getting started because there are practically an endless number of possible combinations and changes in chemical composition.

Additive manufacturing (AM) is a process that creates objects by adding material layer by layer, also known as 3D printing [5]. AM allows for the production of complex geometries and customized parts, which is difficult with traditional manufacturing methods. AM has advantages

Content from this work may be used under the terms of the Creative Commons Attribution 3.0 license. Any further distribution of this work must maintain attribution to the author(s) and the title of the work, journal citation and DOI. Published under license by Materials Research Forum LLC.

Materials Joining and Manufacturing Processes: MJMP 2025   Materials Research Forum LLC
Materials Research Proceedings 55 (2025) 1-6   https://doi.org/10.21741/9781644903612-1

such as low production cost, high precision, and reduced material wastage [6]. The technology is rapidly evolving and has the potential to revolutionize manufacturing processes. Interest in WAAM has increased in recent years. It is a promising alternative to subtractive manufacturing, as it can produce parts with mechanical properties similar to wrought or cast materials. WAAM is a process of direct energy deposition method under AM process, where metal wire is used as deposition material. WAAM relies on the controlled application of heat and deposition of molten metal wire, enabling the layer by layer build-up of metal structures with high precision, strength, and dimensional accuracy [7-8]. WAAM has an ability to create large parts with complex geometry in the most cost-effective manner, as it uses relatively inexpensive materials and equipment than any other traditional methods, therefore it is under consideration in various manufacturing industries [9]. But, WAAM also faces challenges related to inconsistent material properties, questionable structural integrity and surface finish [10]. Post-processing steps like surface modification and post-heat treatment are also required when using Co-Cr alloys in the human body in order to increase durability, decrease ion element release and strengthen the adhesive bond between the tissue and stems [11]. Machining is one of the most widely used method for surface finishing of the components in the manufacturing industries [12]. Since Co-Cr alloys are categorized as hard-to-cut materials, machining them presents a number of difficulties, including low productivity, short tool life, and poor surface quality [13].

Unfortunately, there are no studies on the post-processing of WAAM produced Co-Cr alloys, thus needs attention. Investigating the impact of face milling on the Co-Cr alloy made using the WAAM process is the aim of the current study. The main aim is to have a better understanding of the impact of milling as well as the milling parameters on surface finish of WAAM produced Co-Cr alloy and tool wear.

Materials and methods
Table 1 displays the composition of the Co-Cr alloy filler wire, which was used in the WAAM experiments of this study. The wire had a diameter of 1.2 mm. Operational plan of the study includes the deposition of Co-Cr wire onto stainless steel substrate through pulse mode. Utilizing a Fronius TPS-320i-MIG/MAG welding machine, the WAAM experiments employed a custom-built four-axis welding machine (Fig. 1 (a)). The Cartesian robot, integrated with the welding machine, was controlled via a Delta-PLC controller. Shielding employed 100% argon gas at a constant flow rate of 20 L/min, maintaining a consistent 10 mm distance between the torch tip and workpiece throughout experiments. For multi-layer manufacturing, a 30-second cooling interval followed each deposition layer. Before the deposition process, the stainless steel substrate underwent surface refinement through polishing with 80-grit size polishing paper, ensuring uniform roughness. Following this, impurities were eliminated by cleaning the substrate with cotton soaked in acetone. Later, the substrates were firmly secured in place on the cartesian robot's bed. The cartesian robot's weld deposition on the substrate is performed, deposition was done with 102A current (Table 2). The deposited single layer width and height are observed to be 5.20 mm and 3.34 mm respectively. The target dimension of the alloy component is shown in Fig. 1 (b). The single layer of Co-Cr alloy component produced from WAAM process (Fig. 1 (c)). The alloy's hardness, as determined by the Vickers microhardness device (Mitutoyo HM200), was 783.5 HV.

2

Table 1 *Composition*

Elements	Mn	Si	C	Fe	W	Cr	Co
wt.%	0.8	1.6	2.3	3.0	11.5	26.5	Balance

Table 2 *WAAM process parameters*

Current	102 A
Voltage	20 V
Shielding gas Argon	20 L/min
Wire speed	4.5 m/min
Deposition speed	0.35 m/min

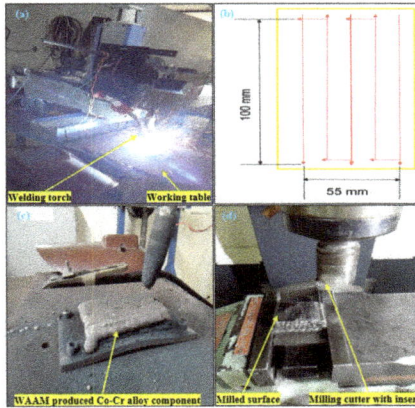

Fig. 1 *(a) WAAM process, (b) Target component dimensions, (c) WAAM produced alloy,*
(d) Milling process

Figure 1 (d) shows the face milling operation on the alloy. The cutter diameter is 50 mm. Initially, the levelling of the work piece was carried out using uncoated carbide insert at a very low cutting speed of 50 m/min. A relatively lower depth of cut (DoC) of 0.1 mm was tested. For the levelling task, the radial DoC was maintained at 50 mm. Once the required level of flatness of the workpiece was achieved, the CVD coated carbide insert is utilized to further proceed with the study of machining conditions suitable for this alloy. Table 3 indicates the cutting conditions used for machining of this alloy using CVD coated insert. The surface roughness (R_a) is measured using 2-D profilometer (Taylor Hobson – Form TalySurf 50). Three readings have been taken on each machined surface, and the average R_a value has been presented in Table 3.

Table 3 *Initial cutting conditions*

v_c (m/min)	Feed rate (mm/z)	Axial DoC (mm)	Radial DoC (mm)	Machining length (mm)	Avg. R_a (μm)
150	0.1	0.1	50	100	0.1266
150	0.15	0.1	50	100	0.2272

Results and discussion

The CVD coated inserts were used for the initial cutting conditions as shown in Table 3. The cutting conditions from Table 3 were tried in order to check the feasibility of the inserts used. During machining, these inserts were not able to withstand the hardness of the alloy and the cutting vibrations. Due to this difficulty, there was a need to modify the cutting conditions. Modification of the machining length from 100 mm to 50 mm and cutting speed from 150 to 50 m/min performed better without damaging the insert. Futher, Bagci et.al. [14] in their study have also recommended PVD coated inserts for face milling of Co-Cr alloy namely stellite 6. During the use of PVD coated insets, it was observed that the heat that arises from tool-chip contact zone will be reflected by coating on the   to the chips. Hence, PVD coated carbide inserts are also tested in comparison to the CVD coated carbide inserts for the cutting conditions presented in Table 4. Figure 2 shows the PVD and CVD coated inserts after single pass.

Table 4 *Investigated cutting conditions*

v_c (m/min)	Feed rate (mm/z)	Axial DoC (mm)	Radial DoC (mm)	Machining length (mm)
50	0.05	0.1	50	50

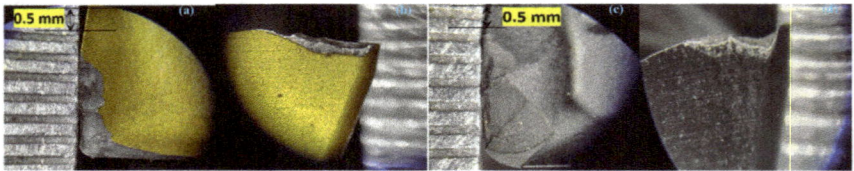

Fig. 2 *(a) Chipping on the rake face (PVD), (b) Chipping on the flank face (PVD), (c) Chipping on the rake face (CVD), (d) Chipping on the flank face (CVD)*

From figures 2 (a), (b), (c), (d), it is observed that the rake face crater width is 2.75 mm for both the CVD and PVD inserts. Here, it is clear that the radial DoC 50 mm was very high for machining since there is chipping of the cutting edge. Hence, the radial DoC was reduced to 5 mm in order to reduce the tool wear. Table 5 represents the cutting conditions which are tested again for these alloys using CVD and PVD inserts.

Table 5 *Recommended machining conditions*

v_c (m/min)	Feed rate (mm/z)	Axial DoC (mm)	Radial DoC (mm)	Machining length (mm)
50	0.05	0.1	5	50

Three machining trials were carried for the machining conditions as shown in Table 5. The flank wear and surface roughness were recorded for both CVD and PVD inserts. Figure 3 shows the flank wear on PVD and CVD inserts, where it is very clear that there was less wear on PVD insert than on CVD insert. Also, machining using PVD coated inserts performed better in terms of producing better surfaces of the alloy with lowered roughness values (Table 6). Both roughness and tool wear results have been presented in Table 6.

Table 6 *The Average R_a and Flank wear for both inserts*

Type of insert	R_a (μm)	Avg. R_a (μm)	Flank wear (mm)	Avg. Flank wear (mm)
PVD	0.0502		0.053	
	0.0514	0.0514	0.065	0.066
	0.0528		0.082	
CVD	0.1714		0.295	
	0.1726	0.1727	0.453	0.462
	0.1740		0.639	

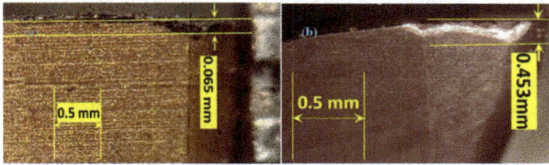

Fig. 3 *(a) Flank wear in PVD insert, (b) Flank wear in CVD insert*

Conclusions

Using the CVD coated carbide and PVD coated carbide inserts to machine wire arc additive manufactured Co-Cr alloy leads to the following important conclusions:

- The machining of WAAM'ed Cobalt-chromium alloy is possible.
- Higher cutting speeds, machining length and radial depth of cut are unsafe to machine these alloys as they lead to tool insert chipping/breakage.
- Shorter machining length, low cutting speed and low radial DoC are recommended to machine these types of alloys.
- PVD coated carbide inserts produce better surface finish of these alloy and reduced tool insert wear, when compared to CVD coated carbide inserts.
- v_c of 50 m/min, feed rate of 0.05 mm/z, axial DoC of 0.1 mm, radial DoC of 5 mm, and machining length of 50 mm is recommended using PVD coated insert to produce better surface roughness (R_a = 0.0514 μm) and reduced tool insert flank wear (0.066 mm).

References

[1] A. Jabbari, Physico-mechanical properties and prosthodontic applications of Co-Cr dental alloys: a review of the literature, J. Adv. Prosthodont. 6 (2014) 138-45. https://doi.org/10.4047/jap.2014.6.2.138

[2] M. Hassim, M. Idris, M. Yajid, S. Samion, Mechanical and wear behaviour of nanostructure TiO2-Ag coating on cobalt chromium alloys by air plasma spray and high velocity oxy-fuel, J. of Materials Research and Technology. 8 (2019) 2290-2299. https://doi.org/10.1016/j.jmrt.2019.04.003

[3] G. Babis, A. Mavrogenis, Cobalt-Chrome Porous-Coated Implant-Bone interface in total joint arthroplasty, In Springer eBooks. (2013) 55-65. https://doi.org/10.1007/978-1-4471-5409-9_5

[4] A. Eissel, L. Engelking, K. Treutler, V. Wesling, D. Schröpfer, T. Kannengießer, Modification of Co-Cr alloys to optimize additively welded microstructures and subsequent surface finishing, Welding in the World. 66 (2022) 2245-2257. https://doi.org/10.1007/s40194-022-01334-0

[5] C. Castillo, F. Saucedo-Zendejo, A. García, Additive Manufacturing Simulation: A Review, In Computational and Experimental Simulations in Engineering. ICCES 2023. Mechanisms and Machine Science. 145 (2024). https://doi.org/10.1007/978-3-031-42987-3_91

[6] A. Järvenpää, D. Kim, K. Mäntyjärvi, Metal additive manufacturing, In Elsevier eBooks. (2023) 493-536. https://doi.org/10.1016/B978-0-323-90552-7.00007-9

[7] K. Vimal, M. Srinivas, S. Rajak, Wire arc additive manufacturing of aluminium alloys: A review, Materials Today: Proceedings. 41 (2021) 1139-1145. https://doi.org/10.1016/j.matpr.2020.09.153

[8] G. Liu, J. Xiong, L. Tang, Microstructure and mechanical properties of 2219 aluminum alloy fabricated by double-electrode gas metal arc additive manufacturing, Additive Manufacturing. 35 (2020) 101375. https://doi.org/10.1016/j.addma.2020.101375

[9] B. Wu, D. Ding, D. Cuiuri, H. Li, J. Xu, J. Norrish, A review of the wire arc additive manufacturing of metals: properties, defects and quality improvement, J. of Manufacturing Processes. 35 (2018) 127-139. https://doi.org/10.1016/j.jmapro.2018.08.001

[10] S. Tofail, E. Koumoulos, A. Bandyopadhyay, S. Bose, L. O'Donoghue, C. Charitidis, Additive manufacturing: scientific and technological challenges, market uptake and opportunities, Materials Today. 21 (2018) 22-37. https://doi.org/10.1016/j.mattod.2017.07.001

[11] C. Un, & P. Spa, Selective laser melting and post-processing stages for enhancing the material behavior of cobalt-chromium alloy in total hip replacement: a review, Materials and Manufacturing Processes. 38 (2022) 495-515. https://doi.org/10.1080/10426914.2022.2105879

[12] V. Marakini, S. Pai, U. Bhat, D. Thakur, B. Achar, Enhancing the surface integrity characteristics of Al-Li alloy using face milling, Materials Letters. 324 (2022) 132610. https://doi.org/10.1016/j.matlet.2022.132610

[13] M. Patar, M. Suhaimi, S. Sharif, A. Mohruni, M. Hisam, M. Shaharum, Influence of machining parameters and tool geometry on tool wear during cobalt chromium-molybdenum micro drilling, Jurnal Teknologi. 86 (2024) 101-114. https://doi.org/10.11113/jurnalteknologi.v86.20842

[14] Bagci, E., & Aykut, Ş. (2005). A study of Taguchi optimization method for identifying optimum surface roughness in CNC face milling of cobalt-based alloy (stellite 6). The International J. of Advanced Manufacturing Technology, 29(9-10), 940-947. https://doi.org/10.1007/s00170-005-2616-y

Materials Joining and Manufacturing Processes: MJMP 2025
Materials Research Proceedings 55 (2025) 7-11

Materials Research Forum LLC
https://doi.org/10.21741/9781644903612-2

Effect of heat treatment on mechanical and wear properties of aluminum tin (Pb free) bearing material

Bhat Jayarama[1,2,a*], Satyanarayan[1,b], B.H. Vadavadagi[2,c],
P.S. Shivakumar Gouda[3,d] and Algur Veerabhadrappa[4,e]

[1]Department of Mechanical Engineering, Alva's Institute of Engineering and Technology Karnataka, Moodbidri, Mangalore 574225 and affiliated to Visvesvaraya Technological University, Belagavi, Karnataka, India

[2]Department of Mechanical Engineering, SDM College of Engineering and Technology, Dharwad -580002 and affiliated to Visvesvaraya Technological University, Belagavi, Karnataka, India

[3]Department of Robotics and AI, Mangalore Institute of Technology & Engineering, Moodbidri, Mangalore 574225 and affiliated to Visvesvaraya Technological University, Belagavi, Karnataka, India

[4]Department of Mechanical Engineering, Rao Bahadur Y Mahabaleshwarappa Engineering College, Ballari 583104, Karnataka, and affiliated to Visvesvaraya Technological University, Belagavi, Karnataka, India

[a]jmb.bhat@gmail.com, [b]satyan.nitk@gmail.com, [c]vbasavaraj99 @gmail.com, [d]ursshivu@gmail.com, [e]veereshalgur@gmail.com

Keywords: Al-15Sn, Pb Free Bearing Material, Alloying Element, Heat Treatment, Quenching, Microstructure, Mechanical, Wear Properties

Abstract. Aluminum tin is a lead-free bearing material and is widely used in automobile industries because of its anti-frictional, self-lubricating property. The present study focused on assessing the mechanical and wear properties of custom-made binary Al-15Sn subjected to heat treatment at 250°C for 1 hour and then cooled in different mediums (furnace, air, and oil). Air-cooled samples showed slightly higher hardness properties, 40% more impact energy, and reduced weight loss (0.2 gms) than furnace (0.32 gms) and oil-cooled (0.52 gms) samples.

Introduction

Babbitt materials have a long history of use in tribological application, due to their excellent strength and surface properties, particularly as bearing material. However, Babitt material contains lead (Pb); since lead is a toxic material, it is restricted from engineering applications. Researchers are focusing on lead (Pb)-free bearing materials and aluminum alloys as alternatives to Babbit materials [1]. Aluminum (Al)-tin(Sn) alloys can function under boundary and dry friction conditions due to the presence of soft tin material, which acts as a solid lubricant; therefore, Al-Sn alloys are also known as self-lubricating materials and are widely used in engine bearing components [2]. However, the mechanical strength of Al-Sn alloys needs to be enhanced to meet modern engine requirements [3]. Researchers have focused on the addition of alloying elements and different heat treatment techniques to improve the properties of the alloy. Bhat et. al. [4] studied the influence of cooling media on the microstructure, impact, and hardness properties of Al–15Sn alloy. Heat-treated samples were quenched in various media, including normal water, cold water, ice, air, and a furnace, to analyze the effects of cooling rates on material properties. The results revealed that the cooling medium significantly impacts the distribution of Sn within the Al matrix. It was mentioned that an enhancement in the alloy's mechanical strength was attributed to the reduction in Sn network thickness and rapid solidification. Jayaram et al. [5]

Content from this work may be used under the terms of the Creative Commons Attribution 3.0 license. Any further distribution of this work must maintain attribution to the author(s) and the title of the work, journal citation and DOI. Published under license by Materials Research Forum LLC.

Materials Joining and Manufacturing Processes: MJMP 2025   Materials Research Forum LLC
Materials Research Proceedings 55 (2025) 7-11   https://doi.org/10.21741/9781644903612-2

reviewed the comparative effects of alloying elements and heat treatment on the properties of Al-Sn alloys. The researchers concluded that heat treatment improves mechanical properties, whereas the alloying element enhances wear properties. Xian-jin et al. [6] focused on the effects of annealing on Al-Sn binary alloy coatings prepared using gas-atomized powders via cold spray processes. Al-5Sn coatings were successfully deposited using high-pressure cold spray with nitrogen, while Al-10Sn coatings required a low-pressure cold spray with helium. Both coatings exhibited dense structures with Sn content consistent with the feedstock powders. Annealing above 200°C led to coarsening and/or migration of the Sn phase, resulting in a significant decrease in microhardness at 250°C. Energy Dispersive X-ray Analysis (EDXA) analysis revealed no significant effect of heat treatment on the Sn phase fraction in Al-5Sn coatings. The bond strength of as-sprayed Al-10Sn coatings was slightly higher than Al-5Sn coatings, while annealing at 200°C improved the bond strength of Al-5Sn coatings. The study by Song et al. [7] on Al–Sn–Si alloys demonstrated significant improvements in mechanical and wear properties through the development of a dual-scale microstructure with a combination of coarse and ultrafine grains. Optimal annealing temperatures (500–550°C) resulted in a balance of strength (205-194 MPa) and ductility (4.9-6.7%) with an enhancement in wear resistance by reducing wear loss and friction coefficients. The formation of an oxide tribolayer during wear testing further improved wear resistance. Schouwenaars et al. [8] explored the mechanical properties and microstructural evolution of a cold-rolled Al-20%Sn-1%Cu alloy subjected to various heat treatments. Optimal mechanical properties were achieved between 300°C and 400°C, demonstrating a simpler approach compared to traditional methods. The study highlights the crucial role of physical phenomena such as precipitation, recovery, and recrystallization during heat treatment. Lower temperatures primarily influenced recovery and precipitation, leading to improved ductility but reduced strength. Conversely, higher temperatures facilitated recrystallization and grain growth, impacting both strength and ductility. Notably, the Ultimate Tensile Strength (UTS) decreased significantly within the first five minutes of annealing at elevated temperatures, while ductility increased, revealing a complex interplay between these mechanical properties. The present study focuses on the effect of heat treatment on aluminum-tin (Al-Sn) alloy with different quenching mediums, which have not been reported by the researchers so far.

Methodology
Custom-made Al-15Sn alloys were prepared by adopting the sand cast route at the foundry by melting pure 85% aluminum and 15% tin ingots in weight percentages using the melting furnace. The cast alloy was then machined to prepare the Charpy impact test specimens as per the ASTM E23 standard (10*10*55 mm) and wear samples as per the G99 standard(Ø8*30mm). The samples prepared were then subjected to heat treatment at 250°C for 1 hour using a muffle furnace, and then samples were allowed to cool in three different mediums, i.e., furnace, air, and oil cooling (Hydraulic quenching oil). The Charpy impact test was carried out on cooled samples using a digital impact testing machine (Daksh Quality Systems PVT. Ltd.). The hardness testing was conducted on the samples using a Rockwell hardness testing machine (Aditya Engineering Service) for aluminum alloys. The B scale was used with a steel ball indenter with a minor load of 10 kg and a 90 kg major load dwelling time of 25 seconds. The wear test was carried out using a pin on the disc machine (Magnum Engineers) at a 50N load, 400RPM, and a track radius of 70mm, and the test was carried out for a duration of 12 minutes. For the microstructure study, the samples were polished using emery paper [P200,P350,P400,P600,P800,P1000,P1500,P2000 grade number] further polished samples were etched with Keller's reagent, and images were studied from the JEOL SEM machine at KUD University, Dharwad.

Materials Joining and Manufacturing Processes: MJMP 2025   Materials Research Forum LLC
Materials Research Proceedings 55 (2025) 7-11   https://doi.org/10.21741/9781644903612-2

Results and Discussion

Scanning electron microscope (SEM) images of alloy-cooled in different media are shown in Figures 1a, 1b, 1c and 1d. As-casted Al–Sn alloy exhibited a fine distribution of tin network (white region) within the aluminium matrix a dark region (Figure 1a). Moreover, the Sn grains appeared to have smooth and uniform morphology. SEM image of Al–Sn alloy solidified within the furnace is shown in Figure 1b. The microstructure shows a non–uniform (random) distribution of Sn morphology. The grain boundaries appeared to be distinct from the as-casted alloy. Moreover, white tin phases appeared to be thicker along the grain boundaries. α–Al phase was found to be in the majority than β–Sn phase.

SEM image of normalised (air-cooled) alloy showed a larger network with thicker β–Sn phases when compared to as-received and furnace-cooled alloys. Sn grains were found to be connected to each other along the grain boundaries of the Al phase, whereas the alloy allowed to cool in hydraulic oil exhibited a coarse structure with larger grain boundaries of α–Al, with β–Sn phases appeared to be thinner and randomly distributed Figure 1d. It indicates that cooling mode has a significant effect on the grain morphology [9, 10, 11]. It was observed that no precipitation of second phase was noticed this is because Al does not alloy readily with solders at temperatures as low as the other metal require [12]

Hardness, impact energy, and wear properties of Al–Sn cooled in different media are presented in Figures 2 and 3 respectively. The residual stress generated during the machining process increased the material's hardness of the as-casted alloy whereas its impact energy and weight loss were found be lower. Slow cooling within the furnace allows the alloy for a significant diffusion and phase equilibrium, resulting in higher hardness, lower impact energy, and a slightly higher weight loss as compared to the as-cast alloy. Oil quenching, being a more rapid cooling method, restricted the time available for α–Al and β–Sn phases to form and grow, leading to the lowest hardness, impact energy, and higher weight loss compared to other samples under consideration. Alloy cooled in the air led to the balance between cooling and retained solid solution strengthening of alloy, due to which the sample designated increased hardness, impact properties, and lower weight loss compared to furnace and oil-quenched samples.

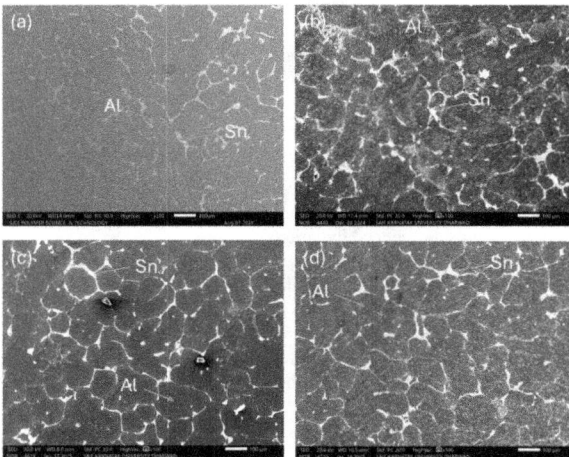

Figure.1 (a) As cast , (b) Furnace cooled ,(c) Air cooled (d) Oil quenched SEM images of Al-15Sn alloy

Materials Joining and Manufacturing Processes: MJMP 2025   Materials Research Forum LLC
Materials Research Proceedings 55 (2025) 7-11   https://doi.org/10.21741/9781644903612-2

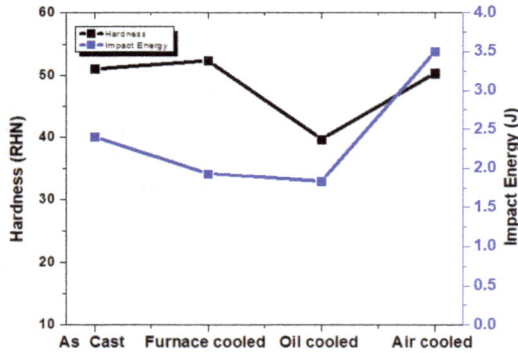

Figure.2 Hardness, Impact energy vs Al-15Sn samples cooled in different medium

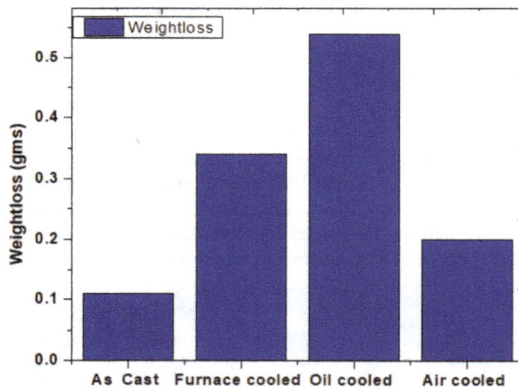

Figure.3 Weight loss vs Al-15Sn alloy cooled in different medium

Conclusion

Based on the results and discussion the following points are concluded:

- Cooling rate significantly influenced the microstructure.
- As– cast and air-cooled samples revealed the smoother microstructure whereas furnace-cooled and oil quenched samples show coarse microstructure.
- A general trend of increased hardness with a decrease in impact energy was observed. Al-15Sn alloy cooled in air appears to be the most effective method to enhance hardness, impact energy and wear properties after heat treatment at 250°C for 1 hour.
- Fine precipitation and solid solution strengthening contributed to the enhanced properties.

References

[1] EJ Abed, Study Microstructure and Mechanical Properties of Rapidly Solidified of Al-Sn By Melt Spinning, Int.Jour.Mech.Eng.15(4) (2015) 53-61.

[2] T Marrocco, LC Driver, SJ Harris, DG McCartney, Microstructure and properties of thermally sprayed Al-Sn-based alloys for plain bearing applications, Jourl. Ther. Spr. Tech. 15 (2006) 634-639. https://doi.org/10.1361/105996306X147009

[3] ZC Lu, Y Gao, MQ Zeng, M Zhu, Improving wear performance of dual-scale Al-Sn alloys: the role of Mg addition in enhancing Sn distribution and tribolayer stability, wear.309 (1-2) (2014) 216-25. https://doi.org/10.1016/j.wear.2013.11.018

[4] J Bhat, Satyanarayan, Effect of Cooling Medium on Microstructure, Impact and Hardness Properties of Al-15Sn Alloy, Tran. Ind. Inst. Met. 72 (2019) 1941-1947. https://doi.org/10.1007/s12666-019-01670-8

[5] J Bhat, R Pinto, Satyanarayan, A review on effect of alloying elements and heat treatment on properties of Al– Sn alloy, Mat.Today: Proce.35 (2021) 340-343. https://doi.org/10.1016/j.matpr.2020.01.617

[6] XJ Ning, JH Kim, HJ Kim, C Lee, Characteristics and heat treatment of cold-sprayed Al-Sn binary alloy coatings, App. Surf.Scie. 255(7) (2009) 3933-3939. https://doi.org/10.1016/j.apsusc.2008.10.074

[7] KQ Song, ZC Lu, M Zhu, RZ Hu, MQ Zeng, A remarkable enhancement of mechanical and wear properties by creating a dual-scale structure in an Al-Sn-Si alloy, Surf. Coat.Tech. 325 (2017) 682-688. https://doi.org/10.1016/j.surfcoat.2017.07.030

[8] R Schouwenaars, JA Torres, VH Jacobo, A Ortiz, Tailoring the mechanical properties of Al-Sn-alloys for tribological applications, InMater. Scie. Foru. 539 9 (2007) 317-322. https://doi.org/10.4028/www.scientific.net/MSF.539-543.317

[9] KN Prabhu, P Deshapande, Satyanarayan, Effect of cooling rate during solidification of Sn-9Zn lead-free solder alloy on its microstructure, tensile strength and ductile-brittle transition temperature, Mater. Scie.Eng. 553 (2012) 64-70. https://doi.org/10.1016/j.msea.2011.11.035

[10]. Satyanarayan, S Tanaka, A Mori, K Hokomoto, Welding of Sn and Cu plates using controlled underwater shock wave,Jour. Mater.Proc.Tech. 245 (2017) 300-308. https://doi.org/10.1016/j.jmatprotec.2017.02.030

[11] Satyanarayan, A Mori, M Nishi, K Hokamoto, Underwater shock wave weldability window for Sn-Cu plates, Jour. Mater.Proc.Tech. 267 (2019) 152-158. https://doi.org/10.1016/j.jmatprotec.2018.11.044

[12] Satyanarayan, KN Prabhu,Wetting behaviour and interfacial microstructure of Sn-Ag-Zn solder alloys on nickel coated aluminium substrates,Mat.Scie.Tech. 27(7)(2011) 1157-1162. https://doi.org/10.1179/026708310X12815992418337

Materials Joining and Manufacturing Processes: MJMP 2025   Materials Research Forum LLC
Materials Research Proceedings 55 (2025) 12-18   https://doi.org/10.21741/9781644903612-3

Development of internal cooling system in hard turning of annealed HCWCI using carbide tools

A.M. Ravi[1,a*], and S.M. Murigendrappa[2,b]

[1]Department of Mechanical Engineering, DRR Government Polytechnic, Davanagere,577004, Karnataka, India

[2]Department of Mechanical Engineering, National Institute of Technology Karnataka (NITK), Surathkal,Mangalore 575025, India

[a]raviam.gpt@gmail.com, [b]smmnitk@gmail.com

Keywords: High Chrome White Cast Iron (HCWCI), Carbide Tools

Abstract. In today's manufacturing techniques, hard turning became the most rising technique to cut high hard materials like high chrome white cast iron (HCWCI). The use of Cubic boron nitride (CBN), Polycrystalline Cubic boron nitride (pCBN) and Carbide tools are most appropriate to shear the metals but are uneconomical. These materials are cut in dry condition, which results in lower life and can be enhanced by minimizing the tool temperature. The research study exposes effective tool cooling systems that can improve tool life and machinability characteristics, also proper selection of cutting parameters to control the cutting temperature. The main aim of this paper is to develop an ideal cooling system to control tool temperature, thereby minimizing the tool wear rates and cutting forces. In this research work, experiments were conducted on annealed high chrome white cast iron using carbide inserts. Intricate cavities were made on the toolholder body for easy circulation of cold water. Statistical tools were used to carry out the experimentations and its analysis. The result confirms, intercooling system minimizes the cutting forces and tool wear at considerable rates compared to non-intercooling system.

Introduction

In our earlier papers, it was focused mainly on machining modelling of high chrome white cast iron (HCWCI) (as-cast type) processed with or without the aid of preheat. During so, derived optimum machinability characteristics of HCWCI, and tool life of cutting tools (CBN and Carbide tools) at various cutting conditions. In hard turning, annealing and hardening of work before and after the metal cutting is the alternative process. In this paper, investigations have been made on annealed HCWCI instead of high hard as-cast type HCWCI to optimize the machining process. Generally, in hard turning, the tool life plays an important role in order to achieve quality products [1]. However, the major problem encountered is at the tool-work interface where large amount of heat generated. This heat softens the uncut material stock but damages the cutting edges [2]. In order to enhance the tool life, cooling of the tool-tip is the only solution. In conventional cooling system, fluid cools both tool and work cutting surfaces, but lowers metal removal since metal shears easily at higher cutting temperatures. Dry cutting due to non-use of coolants is environmentally friendly, which is an important feature in today's green manufacturing. Internal cooling system apparently another good solution but it is rarely used because of its complex manufacturing. The cooling fluid flowing inside the body of the tool necessarily change its phase due to heat at shear zones. Importantly, a closed system does not consume much coolant fluid in the system [3]. The conventional cooling procedures contaminate the tool, workpiece, machine, chips, and the place of work, causing costly cleanings and disposals. On the other hand, an internal cooling system could solve the above said problems; in addition, it acts as a good heat absorbent. These are the positive aspects of effectiveness of internal cooling system [4].

Content from this work may be used under the terms of the Creative Commons Attribution 3.0 license. Any further distribution of this work must maintain attribution to the author(s) and the title of the work, journal citation and DOI. Published under license by Materials Research Forum LLC.

Materials Joining and Manufacturing Processes: MJMP 2025   Materials Research Forum LLC
Materials Research Proceedings 55 (2025) 12-18   https://doi.org/10.21741/9781644903612-3

Numerous authors have worked on internal cooling systems in hard turning process. Among them, Ravi and Murigendrappa [5] conducted series of experiments on as-cast types HCWCI using CBN tools. In their studies, internal cooling system reduces machining force (up to 23%), flank wear (up to 19%) and crater wear (up to 18%). Dhananchezian et al. [6] have investigated the effect of liquid nitrogen in internal cooling system during turning of high hard Ti-6Al- 4V alloy with PVD TiAlN coated tungsten carbide tools (ISO CNMG 120408 MP1 –KC5010). Their results revealed a considerable reduction in the cutting temperature (64 to 67%), cutting force (43 to 53%) and surface roughness (29 to 33%) compared to wet machining process. Wardle et. al. [7] adopted ANN technique to optimize the machinability parameters in internal cooling system. Kromanis et. al. [8] developed new internal cooling systems for modern manufacturing requirements. In Robinson et. al.[3] research work, heating pipe was used to absorb the heat near the cutting tool (multi-coated carbide insert) during turning of AISI 4340 steel supplying minimum fluid. The authors reveals considerable reduction in cutting temperature (22%), tool wear by 15%, surface roughness by 0.83% and the main cutting force by 2.9% compared to conventional method. In their investigational study, Sancheza et. al. [9] used internal cooling system to turn hard Cr-Ni-Nb-Mn-N austenitic steel where cooling fluid used as R-123 hydro chlorofluorocarbon (HCFC). Their results depict the considerable improvement in tool life compare to conventional cutting fluid method. Siller et. al. [10] designed an internal cooling system to cut C45 using carbide tools and explore significant improvement in tool life.

In a modern manufacturing process, 3D printing techniques are commonly used to generate complex cavities in the tool/tool holder more precisely [6]. In this research work 3D printing (Creality Ender V3 S1 model) techniques were used to generate cavities in the tool and the tool holder. Design of experimental tools were used to optimize the effectiveness of internal cooling system and to minimize the cutting temperature and tool wear in hard turning of annealed HCWCI.

Research Methodology
The work material chosen for this research experiment was annealed high chrome white cast iron in the form of round bar of 62mm diameter. The major chemical compositions of this alloy are; 2.7%C, 0.7%Si, 0.66%Mn, 0.068%S, 0.7%P, 0.5%Cu, 27.93%Cr and 0.5%Mo. The important properties of HCWCI are; Melting point 1260 °C, Density 7.7 gm/cc, Thermal conductivity 13 W/mk, Hardness 351 BHN. Cutting tools used were Mitsubishi make Carbide inserts [11]. Figure 1 shows the layout of the internal cooling system used in this machining process. Internal cavities were made on the tool holder and is almost nearer to the tool cutting edges, so that coolant absorbs maximum heat from the shear zone. Holes of 5mm and 3mm diameter were made on the lateral side of the tool holder for the easy entry and exit of the chilled water. The cooling fluid was a purified water measured below the atmospheric temperature (14 °C) and was pumped continuously from a water container to the inlet. The exit water sent back to the container and this forms a closed internal cooling system. The experimental work followed Taguchi's procedures. The control factors and their range selected for the experiments were; cutting speeds of 55, 88 and 136 m/min, depth of cut of 0.1, 0.2 and 0.3 mm, and feed rate of 0.096, 0.124 and 0.179 mm/rev. For each experimental trials, a new cutting edge was used, and the experiments were repeated twice at each condition to keep the experimental error to a minimum value. Figure 2 shows the formation of cavities on the tool holder. Heat dissipation was controlled by winding thick insulation around water pipe. Tool fixed rigidly to the tool holder mounted on a piezoelectric lathe tool dynamometer (Kistler model 9712B500) to measure machining forces viz. Thrust force F_t, Cutting force F_c and Feed force F_f. Arrangements were made to measure the tool tip temperature by mounting K-type thermocouple inside the groves of the tool. In each trial, tool flank and crater wear were measured using scanning electron microscope (SEM). Depth of cut was measured using digital dial gauge, and machining force was calculated using the Equation.1 based on the recorded thrust force, main cutting force and feed forces measured in Newton (N).

$$F_m = \sqrt{F_t^2 + F_c^2 + F_f^2} \tag{1}$$

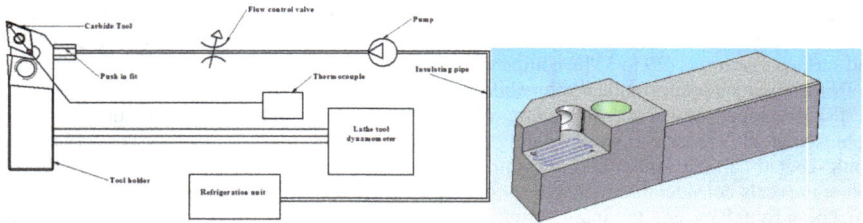

Fig. 1: Layout of Internal cooling system.   Figure 2. Formation of groove on the Tool

Results and Discussions

Table 1 provides the experimental data comprises control factors and process responses viz. machining forces, crater wear, flank wear before and after internal cooling recorded during turning of annealed HCWCI alloy. The complete analysis of these is discussed in the following modules.

Table 1. Taguchi L27(313) array of control factors and process responses

Sl. No.	S [m/s]	d [mm]	f [mm/rev]	F_{m-nc}, [N]	F_{m-c}, [N]	V_{b-nc}, [μm]	V_{b-c}, [μm]	K_{b-nc}, [μm]	K_{b-c}, [μm]	P, [kw]
1	55	0.1	0.096	220	212	18	15	36	29	197
2	88	0.1	0.096	196	168	103	86	200	163	347
3	136	0.1	0.096	152	134	198	166	295	241	648
4	55	0.1	0.124	241	213	17	14	41	33	194
5	88	0.1	0.124	198	168	58	48	212	173	378
6	136	0.1	0.124	155	145	197	166	367	300	643
7	55	0.1	0.179	214	187	18	15	199	162	396
8	88	0.1	0.179	213	198	23	20	264	217	584
9	136	0.1	0.179	213	178	193	162	532	435	643
10	55	0.2	0.096	323	286	24	20	55	45	186
11	88	0.2	0.096	307	270	171	143	127	105	384
12	136	0.2	0.096	270	219	196	165	637	522	598
13	55	0.2	0.124	322	278	22	18	78	64	197
14	88	0.2	0.124	284	241	160	134	174	143	445
15	136	0.2	0.124	231	196	198	166	648	530	561
16	55	0.2	0.179	247	214	23	20	427	349	323
17	88	0.2	0.179	248	213	80	68	250	204	505
18	136	0.2	0.179	252	214	201	168	656	537	517
19	55	0.3	0.096	371	219	40	33	246	201	215
20	88	0.3	0.096	365	305	111	93	266	218	465
21	136	0.3	0.096	347	303	178	150	278	228	509

22	55	0.3	0.124	362	307	47	39	268	219	267
23	88	0.3	0.124	336	286	169	142	269	219	483
24	136	0.3	0.124	284	242	194	164	287	235	499
25	55	0.3	0.179	339	288	102	85	587	481	427
26	88	0.3	0.179	342	291	265	223	273	223	493
27	136	0.3	0.179	350	297	218	183	329	269	495

Machining force: This is due to more volume of work material removal per time at tool-work interface [12]. However, it decreases as the feed rate increases due to enough heat transfer to the stock material. At constant feed rate, as the cutting speed increases machining force decreases. Generation of sufficient amount of cutting temperature at the tool-work interface may be the reason for the same [13].

Similarly, at constant depth of cut, machining force decreases as the cutting speed increases. Figure 3a depicts the characteristic curves of machining forces (Fm-nc) at different cutting speeds and feed rates during dry conditions (without cooling).

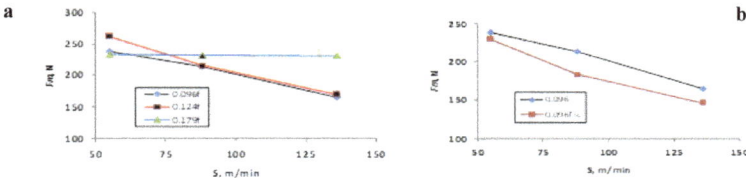

Fig. 3: Characteristic curves of (a) machining force, F_{m-nc} Vs. cutting speeds at a constant depth of cut. (0.2mm) with no cooling (b) machining force F_m V_s cutting speed, S at constant depth of cut 2 mm and feed rate 0.096 mm/rev with and without cooling.

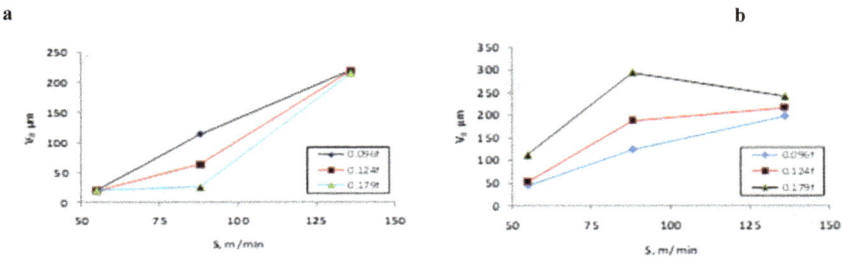

Fig. 4: Characteristic curves of (a) flank wear, V_{b-nc} V_s cutting speed, S at a constant depth of cut (0.1 mm) without cooling (b) flank wear, V_{b-nc} V_s cutting speed, S at a constant depth of cut (0.3 mm) without cooling.

It is understood from the Table1 that internal cooling system is more effective compared to dry machining conditions. The characteristic curves constructed for machining force, Crater wear and flank wear at various cutting conditions with and without internal cooling system exhibits similar trend but there are considerable differences in their values. Sufficient cooling of cutting edge might be the reason for this, thereby tool retain its properties. In Figure 3b characteristic curves shows

Materials Joining and Manufacturing Processes: MJMP 2025   Materials Research Forum LLC
Materials Research Proceedings 55 (2025) 12-18   https://doi.org/10.21741/9781644903612-3

comparative study where machining forces in internal cooling system (F_{m-c}) is well below the machining force generated in dry conditions (F_{m-nc}) at constant feed rates.

Tool wear: Among all six wear mechanisms such as abrasion, adhesion, attrition, fatigue, dissolution/diffusion, and tribochemical wear or any combinations of these are influence on wear of cutting tools [12]. However, abrasion, adhesion, and diffusion mechanisms are more important one. The parameters like tool content, binder phase, chemical stability of tool, and its composition are influence most on tool wear [13]. Interestingly, flank wear is considered as the key wear mechanism in tool life metric in this machining process. The experimental work responses shown in the Table.1 highlights the influence of internal cooling on the flank wear and crater wear. Figures 3.2.1. (a) and (b), shows the characteristic curves of flank wear at different depth of cuts, feed rates and cutting speeds. At lower depth of cut (0.1 mm), and for all feed rates, cutting speeds flank wear curves exhibits linearity (Fig. 4a) compared to higher depth of cuts (0.3mm) (Fig. 4b). Ravi and Murigendrappaet al. [12] discussed in hard turning, at higher depth of cuts, volumetric metal removal rate is high which carries enough amount of heat developed at the shear zone. Whenever hot chip flows on the rake face of the tool due to adhesion and abrasion phenomenon stick to flank face of the tool and damages severely. Figures 5a and 5b show the comparative study of flank wear with (Vb-c) and without (Vb-nc) internal cooling system for 0.096 and 0.179 feed rates.

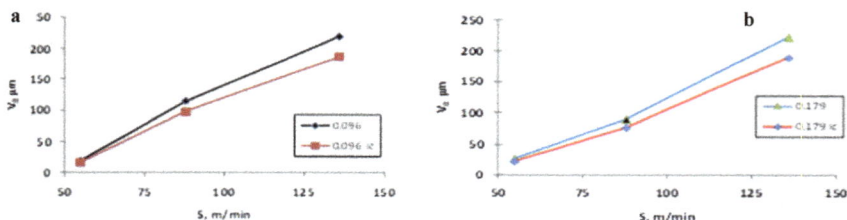

Fig. 5a: Characteristic curves of (a) flank wear, V_{b-nc} V_s cutting speed, S at a constant depth of cut (0.1 mm) and feed rate (0.096 mm/rev) with and without cooling (b) flank wear, V_{b-nc} V_s cutting speed, S at a constant depth of cut (0.1 mm) and feed rate (0.179 mm/rev) with and without cooling

It is understood from Fig. 5a and 5b that, at a constant depth of cut and feed rate, as the cutting speed increases flank wear increases. However, wear rate is minimum to considerable extent due to internal cooling. When hot chip flows on the tool rake face, the coolant circulated inside the tool holder observes the maximum heat there by tool retain its properties. The characteristic curves follow similar trajectory but are differs in their values for the same control parameters. Interestingly, like flank wear crater wear also occurs due to abrasion and diffusion of molten chip on the tool crater surface [12].

During the machining process, the cut metal (chip) slides on the rake face of the tool carries huge amount of heat in which major portion is transfer to the tool. This causes considerable changes in tool thermal and mechanical properties. Further, this leads to more diffusion wear on the surface [13]. In order to retain these properties and to enhance tool life cooling is the best solution. The characteristics curves shown in Figs. 6a and 6b describes the behavior of crater wear with cooling (Kb-c) and with no-cooling system (Kb-nc) for different cutting parameters.

Materials Joining and Manufacturing Processes: MJMP 2025   Materials Research Forum LLC
Materials Research Proceedings 55 (2025) 12-18   https://doi.org/10.21741/9781644903612-3

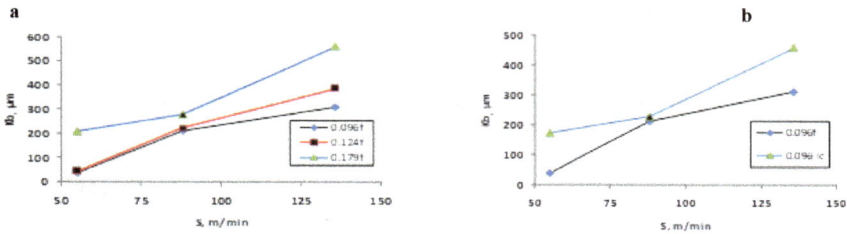

Fig. 6: Characteristic curves of (a) crater wear, K_b V_s cutting speed, S at a constant depth of cut (0.1 mm), and feed rates (0.096, 0.124 and 0.179 mm/rev) with no cooling (b) K_b V_s cutting speed, S at a constant depth of cut (0.1 mm) and feed rate (0.096 mm/rev) with and without cooling.

At constant depth of cut, crater wear increases with increase of cutting speed. The maximum wear occurs at lower cutting speeds, which leads to the formation of break-up chips and good surface finish. Almost similar results were obtained even at medium cutting speeds, however higher cutting speeds generates smooth surface finish. Fig. 6b shows the characteristics curves where internal cooling is more effective than dry machining process. In most of the cases annealed HCWCI exhibits best compared to high hard as-cast type HCWCI alloy.

Conclusion
In this new research experimental studies, machinability characteristics of annealed HCWCI alloy exhibits best results with internal cooling system compared dry machining process. Also, research analysis stated that Carbide tools can be an alternative to CBN tools. Newly designed internal cooling system looks more economical and effective to enhancing tool life. Importantly, use of carbide tools optimizes tool cost considerably in part machining. Adopted Taguchi's procedure to optimize the experimental work and helps to analyze the influence of control parameters on machining of annealed HCWI and tool wear. The experimental results confirm considerable decrement in machining force (up to 18%), flank wear (up to 17%) and crater wear (up to 18%) in internal cooling system as compared to dry machining process. Considering the economics of metal cutting process, internal cooling system yields better results compared to dry machining process.

References

[1] T. Kitagawa, A. Kubo, K. Maekawa, Temperature and wear of cutting tools in high-speed machining of Inconel 718 and Ti-6Al-6V-2Sn, Wear 202 (1997) 142-148. https://doi.org/10.1016/S0043-1648(96)07255-9

[2] Y. Isik, An experimental investigation on effect of cutting fluids in turning with coated carbides tool, StrojVestn: JME 56 (2010) 252-262.

[3] R. R. Gnanadurai, A.S. Varadarajan, Investigation on the effect of cooling of the tool using heat pipe during hard turning with minimal fluid application, IJEST (2016).

[4] A. Siller, Steininger, F. Bleicher, Heat dissipation in turning operations by means of internal cooling, Procedia Engineering 100 (2015) 1116-1123. https://doi.org/10.1016/j.proeng.2015.01.474

[5] A. M. Ravi, S. M. Murigendrappa, Experimental study on internal cooling system in hard turning of HCWCI using CBN tools, AIP Conference Proceedings 1943 (2018). https://doi.org/10.1063/1.5029629

[6] M. Dhananchezian, D. Satishkumar, S. Palani, N. Ramprakash, Study the effect of cryogenic cooling with modified cutting tool insert in the turning of Ti-6Al-4V alloy, International Journal of Engineering Research & Technology (IJERT) 2(9) (2013) 2278-0181.

[7] F. Wardle, T. Minton, S. Bin Che Ghani, P. Förstmann, M. Roeder, S. Richarz, F. Sammler, Artificial neural networks for controlling the temperature of internally cooled turning tools, Modern Mechanical Engineering 3 (2013) 1-10. https://doi.org/10.4236/mme.2013.32A001

[8] G. Kromanis,. G. Pikurs, K. Muiznieks, V. Kravalis, Gutakovskis, Design of internally cooled tools for dry cutting, 9th International DAAAM Baltic Conference "Industrial Engineering", Tallinn, Estonia, April 24-26, 2014.

[9] L. E. A. Sanchez, V. L. Scalona, G. G. C. Abreu, Cleaner machining through a toolholder with internal cooling, 3rd Workshop on Cleaner Production Initiatives and Challenges for a Sustainable World, São Paulo, Brazil, May 18-20, 2011.

[10] Siller, F. Steininger, Bleicher, Heat dissipation in turning operations by means of internal cooling, 25th DAAAM International Symposium (2015).

[11] Mitsubishi Materials Catalogue (2011-12).

[12] A. M. Ravi, S. M. Murigendrappa, P. G. Mukunda, Machinability investigations on high chrome white cast iron using multi-coated hard carbide tools, TIIM - Springer Publications 67 (2014) 485-502. https://doi.org/10.1007/s12666-013-0369-0

[13] A. M. Ravi, S. M. Murigendrappa, P. G. Mukunda, Experimental investigation on thermally enhanced machining of high-chrome white cast iron and study of its machinability characteristics using Taguchi method and artificial neural network, International Journal of Advanced Manufacturing Technology (IJAMT) 72 (2014) 1439-1454. https://doi.org/10.1007/s00170-014-5752-4

Materials Joining and Manufacturing Processes: MJMP 2025   Materials Research Forum LLC
Materials Research Proceedings 55 (2025) 19-26   https://doi.org/10.21741/9781644903612-4

Investigation of mechanical properties of natural material 3D printed specimens

Vinay Swamy[1,a], Gurupadayya[1,b], Naveen Kumar[1,c], Vishalagoud S. Patil[1,d]*, Gurushant B. Vaggar[2,e], and Dodda Hanamesha[1,f]

[1]Department of Mechanical Engineering, Government Engineering College Talakal Karnataka, Koppal 583238, India

[2]Department of Mechanical Engineering, Alva's Institute of Engineering Technology, Mijar, Moodbidri, Karnataka, India

[a]hvinay50@gmail.com, [b]guruguttedar444@gmail.com, [c]naveenmalekoppa496@gmail.com, [d]vspatil8018@gmail.com, [e]gvaggar7@aiet.org.in, [f]hanumesh.gaddi@gmail.com

Keywords: 3D Printing, Tension, Natural Fiber, Filament, PLA, ABS

Abstract. 3D printing, or additive manufacturing, has indeed revolutionized multiple industries by offering a more efficient, flexible, and precise way to create products. It is the process of building up material layer by layer based on a 3D CAD model, which allows for highly complex geometries that traditional manufacturing methods might not be able to produce. This competence is a key reason for aerospace, automotive, healthcare, and consumer goods sectors have embraced 3D printing technologies. The present study investigates the use of natural materials in 3D printing, aiming to provide a sustainable and eco-friendly alternative to traditional synthetic materials. The research focuses on identifying suitable natural materials, evaluating their mechanical properties by testing UTM and assessing their environmental impact. Key natural materials explored include ABS.

Introduction

3D printing, also known as additive manufacturing, is a technique used to create detailed and complex forms of materials and components by building them layer by layer. It is sometimes referred to as fused filament fabrication (FFF) or fusion deposition modelling (FDM), the process of melting a filament and extruding it through a heated nozzle to form each layer. The filament plays a fundamental role in this process, through which the object is built [1-5]. The filament typically comes in spools and can be made from a variety of substances such as plastic (like PLA, ABS, PETG), metals, ceramics, and even biomaterials. The selection of filaments depends on the application and the properties required for the final product, such as strength, flexibility, heat resistance, or durability.

Filaments used in 3D printing are primarily made from a variety of materials, with polymer-based filaments being the most common [1-5]. These filaments are important for producing durable, flexible, and detailed objects. Each filament type has specific properties that make it suitable for different applications. With many filament options available, 3D printing has become highly versatile, enabling the creation of everything from everyday household items to high-performance industrial components. Choosing the right filament depends on the specific requirements of the project, such as mechanical properties, aesthetics, and printability.

The selection of filament for 3D printing should be based on the specific needs of the application, whether it be strength, flexibility, food safety, biodegradability, or other specialized properties. The versatility of materials like Metal/PLA, Nylon, PETT, and PLA ensure that 3D printing can be used across a wide range of industries, like mechanical engineering and medical devices, food packaging and consumer goods.

Content from this work may be used under the terms of the Creative Commons Attribution 3.0 license. Any further distribution of this work must maintain attribution to the author(s) and the title of the work, journal citation and DOI. Published under license by Materials Research Forum LLC.

Materials Joining and Manufacturing Processes: MJMP 2025   Materials Research Forum LLC
Materials Research Proceedings 55 (2025) 19-26   https://doi.org/10.21741/9781644903612-4

In modern 3D printing, the focus on producing lightweight materials has become very important, especially in industries where weight reduction is crucial for performance, efficiency, and cost-effectiveness. The Fusion Deposition Modelling (FDM) process, a popular 3D printing technique, plays a important role in this effort by enabling the fabrication of complex, lightweight structures using a variety of materials [6-8].

Mechanical properties are the important characteristics found in each material and describe how each material behaves under various forces and stresses. These properties help in critical selection of materials in engineering and design. Some of the important mechanical properties are strength, hardness [9], elasticity [10], plasticity [11], ductility [12], brittleness [13], toughness etc. Each of these properties are affected by changes in temperature, environmental conditions, processing methods and material composition. While designing products, careful selection of materials with the right combination of these properties is very important.

The aim of the present study is to investigate the effect of print layer orientation on the tensile properties of PLA/ABS polymer [14]. Specifically, it will analyze for different printing positions influencing the tensile stress and material strength. PLA is chosen for its environmentally friendly characteristics, including its biodegradability, where the influence of printing direction on mechanical properties like tensile strength is important for optimizing the material's performance[4].

Materials and methods

In Fused Deposition Modelling (FDM) 3D printing [15], the bulk raw material is typically a plastic filament. This filament is a thermoplastic material that is used specifically for additive manufacturing (AM). ABS (Acrylonitrile Butadiene Styrene) is a thermoplastic polymer recognized for its strength, toughness, and impact resistance.

The test specimen as per standard ISO 527 is shown in figure 1 below. The computer aided drawing of standard specimens is prepared in Solid-edge software and is as revealed in figure 2. The dimensions are according to standard ISO 527 as mentioned in figure 1. Then the CAD model was transformed into STL (stereolithography) file. Then the model was manufactured in 3D printer of Anker make M5 model using Ultimaker Cura software [16]. The specimen was printed on desktop FDM 3D printer shown in figure 3 below.

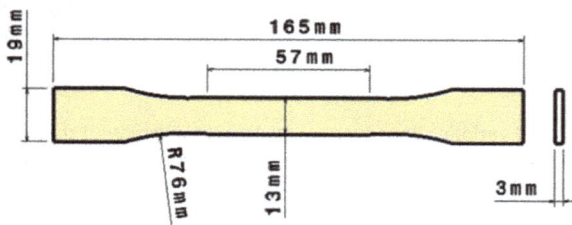

Fig. 1. Standard Test Specimen

Materials Joining and Manufacturing Processes: MJMP 2025

Materials Research Proceedings 55 (2025) 19-26

Materials Research Forum LLC

https://doi.org/10.21741/9781644903612-4

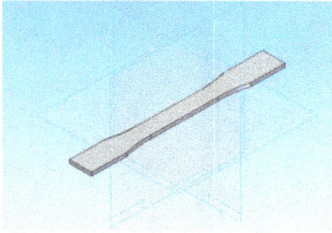

Fig. 2. CAD model as per ISO 527

Fig. 3. 3D printing machine

Experimental Work

Tensile test specimens were manufactured in two orientations, flat and vertical [17,18]. Prior to initiating the printing process the parameters are configured as mentioned in Table 1. The material is then printed in different print layer orientation directions specifically flat and vertical positions illustrated as in figure 4a and 4 b [19].

Table 1. Parameters of Printer settings

Item	Value
Filament	PLA, ABS, ASA
Filament Diameter	1.75 mm
Speed of Tool Head	500 mm/sec
Layer Height	0.3 mm
Print Head Temperature	350 degrees
Bed Temperature	120 degrees
Infill	100 %

Fig. 4 a. Flat orientation

Fig. 4b. Vertical orientation

Tensile Testing

Tensile test was carried out using a micro universal testing machine in accordance with the ASTM D368 [20-21] as shown in figure 5a and 5b. Tests were conducted at constant crosshead speed of 2 mm/m in until failure, with data on stress and strain recorded throughout. The specimen dimension consists of a length of 165 mm and a width of 19 mm in the grip side area, and a width of 57 mm in the gauge length area and material flat of thickness 3 mm.

Materials Joining and Manufacturing Processes: MJMP 2025   Materials Research Forum LLC
Materials Research Proceedings 55 (2025) 19-26   https://doi.org/10.21741/9781644903612-4

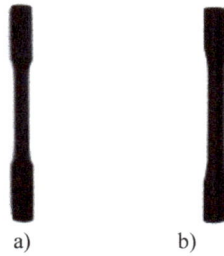

a)   b)

Fig. 5. Actual tensile specimens a) Flat and b) Vertical Samples

The flat was printed directly into the heated substrate which affects the shape of the layer alignments and vertical position specimens were built upright as shown in figure 6.

Fig. 6. Tensile test setup

Results and discussions

The stress strain diagram for 3D printed ABS material [22] for flat and vertical orientation are as shown in figure 7. The results indicate that the highest stress is observed in 0° orientation than 90° orientation [23] of 3D printed ABS specimens. These variations in stress are due to changes in print layer orientation and loading directions on the printed materials [24]. The stresses developed in both the orientation specimens are mentioned in Table 2.

Materials Joining and Manufacturing Processes: MJMP 2025   Materials Research Forum LLC
Materials Research Proceedings 55 (2025) 19-26   https://doi.org/10.21741/9781644903612-4

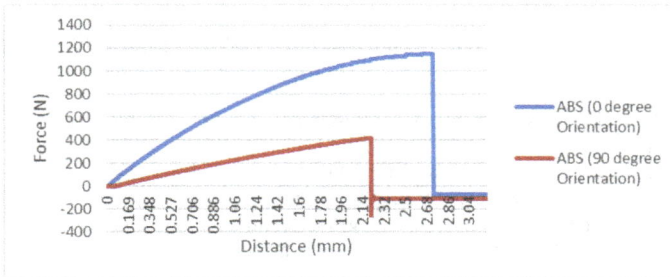

Fig. 7. Stress strain diagram of ABS material

Table 2. Tensile Properties

No	Printing position	Stress (MPa)	Strain (%)	Force (N)	Distance (mm)
1	0	20	6.2	1150	2.7
2	90	09	0.6	413	2.2

In relation to mechanical properties, it can be realized from Figure 7 and Table 2 that the flat position sample reaches maximum tensile strength (1150 N), while the vertical position reaches the minimum (413 N). Because the tensile strength of FDM 3D-printed ABS parts becomes optimal if the parts are fully oriented along the path of loading stress [25-28].
Figure 8 shows the broken specimen and failure mechanism in each specimen printed with different directions [29-30].

Fig. 8. Broken ABS specimens at two different orientations

The vertical sample exhibits a brittle fracture with striated cracking, as depicted in Figure 8. In the horizontal sample, bending deformation occurs along a straight member. This causes contraction in the fibers above the neutral axis and elongation in the fibers below the neutral axis.

Conclusion

In this paper, the mechanical tensile properties of the FDM 3D printing technology have been examined for two different built positions. The following conclusions were drawn from the mentioned study.

- The ABS 3D specimen was successfully manufactured in accordance with ASTM standards.
- The highest tensile stress is observed in materials printed with flat position.
- The maximum force that the sample withstand are determined to be 1150 N for materials printed at flat position. Conversely, the minimum force that the sample withstand is recorded as 413 N for materials printed in an vertical position.

References

[1] D Zhuang, Z Ning, Y Chen, J Li, Q Li, W Xu, Investigation on mechanical properties regulation of rock-like specimens based on 3D printing and similarity quantification, Int. J. Mining Sci. and Tech 34, (2024), 573-585. https://doi.org/10.1016/j.ijmst.2024.05.004

[2] V.D. Sagiasa, K.I. Giannakopoulosa, C. Stergioua, Mechanical properties of 3D printed polymer specimens, 1st Int. Conf of the Greek Society of Experimental Mechanics of Materials, (2018), 85-90. https://doi.org/10.1016/j.prostr.2018.09.013

[3] D Pezer, F Vukas, M Butir, Experimental Study of Tensile Strength for 3D Printed Specimens of HI-PLA Polymer Material on In-house Tensile Test Machine, Technium Vol. 4, No.10 pp.197-206 (2022), www.techniumscience.com

[4] A Alhuzaim, Investigating The Mechanical Properties of Pla Polymer Tensile Test Samples Produced Via 3D Printing in Various Orientations: Flat, Vertical, and 45 Degrees, Inter J of Mecha Engg and Techn (IJMET) Volume 15, Issue 2,March-April(2024), pp. 35-42, https://iaeme.com/Home/issue/IJMET?Volume=15&Issue=2

[5] S Brischetto and R Torre, Tensile and Compressive Behavior in the Experimental Tests for PLA Specimens Produced via Fused Deposition Modelling Technique, J. of Composite Sci, (2020), 1-25. https://doi.org/10.3390/jcs4030140

[6] T Appalsamy, S L Hamilton and M J Kgaphola, Tensile Test Analysis of 3D Printed Specimens with Varying Print Orientation and Infill Density, J. of Composite Sci, (2024), 8, 121. https://doi.org/10.3390/jcs8040121

[7] J Sedlak, Z Joska, J Jansky, J Zouhar, A and J Jirousek, Analysis of the Mechanical Properties of 3D-Printed Plastic Samples Subjected to Selected Degradation Effects, Materials (2023), 16, 3268. https://doi.org/10.3390/ma16083268

[8] T M. Lazovi, M S. Stojanovi, Preparation of specimens for standard tensile testing of plastic materials for FDM 3D printing, Int. Sci Journal Machines, Technologies. Materials Year XV, Issue 5 , PP. 205-208 (2021). https://stumejournals.com/journals/mtm/2021/5/205

[9] Fattepur, G., Patil, A.Y., Kumar, P. *et al.* Bio-inspired designs: leveraging biological brilliance in mechanical engineering an overview. *3 Biotech* **14**, 312 (2024). https://doi.org/10.1007/s13205-024-04153-w

[10] Dhaduti, S.C.; Sarganachari, S.G.; Patil, A.Y.; Budapanahalli, S.H.; Kumar, R. Asymmetric/Symmetric Glass-Fibre-Filled Polyamide 66 Gears—A Systematic Fatigue Life Study. J. Compos. Sci. 2023, 7, 345. https://doi.org/10.3390/jcs7090345

[11] A Dharmdas, A Y. Patil, A Baig, O Z. Hosmani, S N. Mathad, M B. Patil, R Kumar, B B. Kotturshettar and I Md R Fattah, An Experimental and Simulation Study of the Active

Camber Morphing Concept on Airfoils Using Bio-Inspired Structures, Biomimetics 2023, 8(2), 251; https://doi.org/10.3390/biomimetics8020251

[12] P A Y., M T H.M., K Akshay B., M Shridhar N., Patil M B., Thermo gravimetric analysis study of kinematic parameters and statistical analysis for big sheep horn/scapula bone of Indian origin, Acta Periodica Technologica 2023 Issue 54, Pages: 21-35.

[13] S Karadgi*, P M. Bhovi, A Y. Patil, K Ramaiah, K. Venkateswarlu and T G. Langdon, A Conceptual Framework Towards the Realization of in situ Monitoring and Control of End-to-End Additive Manufacturing Process, Micro and Nanosystems, Volume 15, Issue 2, 2023, Published on: 09 June, 2023, Page: [92 - 101]. https://doi.org/10.2174/1876402915666230405132640

[14] A Y. Patil, C Hegde, G Savanur, S Mohammed Kanakmood, A M. Contractor, V B. Shirashyad, R M. Chivate, B B. Kotturshettar, S N Mathad, M B Patil, M E M Soudagar, I. M. R. Fattah, Biomimicking the nature-inspired design structures – an experimental and simulation approach with aid of additive manufacturing, Biomimetics, MDPI, Accepted, https://doi.org/10.3390/biomimetics7040186

[15] Choukimath, M.C.; Banapurmath, N.R.; Riaz, F.; Patil, A.Y. Jalawadi, A.R. Mujtaba, M.A. Shahapurkar, K. Khan, T.M.Y. Alsehli, M.; Soudagar, M.E.M.; Fattah, I.M.R. Experimental and Computational Study of Mechanical and Thermal Characteristics of h-BN and GNP Infused Polymer Composites for Elevated Temperature Applications. Materials 2022, 15, 5397. https://doi.org/10.3390/ma15155397

[16] Budapanahalli, S.H. Mallur, S.B. Patil, A.Y. Alosaimi, A.M.; Khan, A. Hussein, M.A. Asiri, A.M. A Tribological Study on the Effect of Reinforcing SiC and Al_2O_3 in Al7075: Applications for Spur Gears. Metals (2022), 12, 1028. https://doi.org/10.3390/met12061028

[17] Shankar A Hallad, N R Banapurmath, Avinash Bhadrakali, A Y. Patil, A M Hunashyal, S V Ganachari, T M Yunuskhan, I A Badruddin, M elahi M S., Sa Kamangar, Nanoceramic Composites for Nuclear Radiation Attenuation, Materials (2022), 15(1), 262; https://doi.org/10.3390/ma15010262

[18] K, M.H. Kumar, M.P. Patil, A.Y. Keshavamurthy, R. Khan, T.M.Y. Badruddin, I.A. Kamangar, S. Analysis of the Effect of Parameters on Fracture Toughness of Hemp Fiber Reinforced Hybrid Composites Using the ANOVA Method Polymers (2021), 13, 3013. https://doi.org/10.3390/polym13173013

[19] Keshavamurthy R.; Tambrallimath V., Rajhi. A. A. M. S. A. R, Patil, A. Y. Khan, Y. T. M. Makannavar, R.; Influence of solid lubricant addition on friction and wear response of 3D printed polymer composites. Polymers (2021), 13. https://doi.org/10.3390/polym13172905

[20] Tambrallimath V.; Keshavamurthy R.; Bavan D. S., Patil, A. Y., Khan, Y. T. M, Badruddin, I. A.; Kamangar, S, Mechanical properties of PC-ABS based graphene reinforced polymer nano composite fabricated by FDM process. Polymers (2021), 13, https://doi.org/10.3390/polym13172951

[21] Mysore, T.H.M.; Patil, A.Y. Raju, G.U. Banapurmath, N.R. Bhovi, P.M. Afzal, A.; Alamri, S.; Saleel, C.A. Investigation of Mechanical and Physical Properties of Big Sheep Horn as an Alternative Biomaterial for Structural Applications. Materials (2021), 14, x. https://doi.org/10.3390/ma14144039.

[22] N. Vijaya Kumar, N. R. Banapurmath, A M. Sajjan, A Y. Patil and Sharanabasava V Ganachari, Studies on Hybrid Bio-nanocomposites for Structural applications, Journal of Materials Engineering and Performance, (2021). https://doi.org/10.1007/s11665-021-05843-9

[23] V S. Patil, Farheen Banoo, R.V. Kurahatti, G.U. Raju, Manzoore Elahi M. Soudagar, Ravinder Kumar, C. Ahamed S, A study of sound pressure level (SPL) inside the truck cabin for new acoustic materials: An experimental and FEA approach, Alexandria Engg, J, (2021). https://doi.org/10.1016/j.aej.2021.03.074

[24] Arun Y. Patil, Akash Naik, Bhavik Vakani, Rahul Kundu, N. R. Banapurmath, Roseline M, Lekha Krishnapillai, Shridhar N. Mathad, Next Generation material for dental teeth and denture base material: Limpet Teeth (LT) as an alternative reinforcement in Polymethylmethacrylate (PMMA), JOURNAL OF NANO- AND ELECTRONIC PHYSICS, Sumy State University, Vol. 5 No 4, 04001(7pp) (2021). https://doi.org/10.21272/jnep.13(2).02033

[25] Vijay Tambrallimath, R Keshavamurthy, Arun Y. Patil, Adarsh H, Mechanical and Tribological characteristics of polymer composites developed by fused filament fabrication, Accepted, Book Chapter, Fused Deposition Modeling based 3D Printing, Springer-Nature, (2020)

[26] Sandeep Dhaduti, S. R. Ganachari and Arun Y. Patil, Prediction of injection molding parameters for symmetric spur gear, Journal of Molecular Modeling, Springer Nature, Oct (2020), Springer Nature publications. https://doi.org/10.1007/s00894-020-04560-9

[27] Prabhudev S Yavagal, Pavan A Kulkarni, Nikshep M Patil, Nitilaksh S Salimath, Arun Y. Patil, Rajashekhar S Savad, B B Kotturshettar, Cleaner production of edible straw as replacement for thermoset plastic, Elsevier, Materials Today Proceedings, March (2020). https://doi.org/10.1016/j.matpr.2020.02.667

[28] Shruti Kiran Totla, Arjun M Pillai, M Chetan, Chetan Warad, Arun Y. Patil, B B Kotturshettar, Analysis of Helmet with Coconut Shell as the Outer Layer, Elsevier, Materials Today Proceedings, March (2020). https://doi.org/10.1016/j.matpr.2020.02.047

[29] Arun Y. Patil, N. R. Banapurmath, B B Kotturshettar, Lekha K, Roseline M, Limpet teeth-based polymer nanocomposite: a novel alternative biomaterial for denture base application, Elsevier, Chapter, In book: Fiber-Reinforced Nanocomposites: Fundamentals and Applications, Jan (2020). https://doi.org/10.1016/B978-0-12-819904-6.00022-0.

[30] Alok R. Shivappagoudar, Amit S. Gali, Anirudh V. Kuber, Sadashivu I. Giraddi, Arshad N. Havaldar, Arun Y. Patil, B B Kotturshettar, R Keshavamurthy, Design Optimization of Innovative Foldable Iron Box, Springer-Nature, Chapter, In book: Innovative Product Design and Intelligent Manufacturing Systems, Jan (2020). DOI: 10.1007/978-981-15-2696-1_5

Materials Joining and Manufacturing Processes: MJMP 2025   Materials Research Forum LLC
Materials Research Proceedings 55 (2025) 27-33   https://doi.org/10.21741/9781644903612-5

Effect of inclination angle in explosive welding of magnesium-zinc sheets: A numerical approach

Samuel Debbarma[1,a]*, Subrata K. Ghosh[1,b], S. Saravanan[2,b], Prabhat Kumar[3,b]

[1]Department of Mechanical Engineering, National Institute of Technology, Jirania, Agartala, Tripura, 799046, India

[2]Department of Mechanical Engineering, Annamalai University, Annamalai Nagar, Chidambaram, TamilNadu, 608002, India

[3]Department of Mechanical Engineering, Galgotias University, Greater Noida, Uttar Pradesh-203201, India

[a]samuel24debbarma@gmail.com, [b]subratagh82@gmail.com, [c]ssvcdm@gmail.com, [d]prabhat.kumar1@galgotiasuniversity.edu.in

Keywords: Magnesium (Mg), Zinc (Zn), Biocompatible, Pressure, Velocity, Strain, AUTODYN, Smoothed Particle Hydrodynamics (SPH)

Abstract: Magnesium (Mg) and Zinc (Zn) are among the most biocompatible materials for humans, leading to substantial global research on these metals. The joining of magnesium and zinc are challenging because of their distinct properties (melting point, thermal conductivity, hardness, and strength). This article provides insights into the effect of employing an inclination angle for obtaining a successful joint by explosive welding. A 2D model is developed and simulated using ANSYS AUTODYN to determine the pressure, velocity and strain developed during parallel and inclined explosive welding. The pressure, velocity, and strain obtained during simulation in inclined configuration exceed the parallel configuration, indicating jetting and undulating interface.

Introduction

Implants are the most effective method for addressing surgical intervention for bone fracture diagnosis and treatment, facilitating the regeneration of fractured bone [1]. Implants are normally made of metals such as titanium, stainless steel, and cobalt-chromium and are thoroughly evaluated and applied in medical practice [2]. However, owing to the non-degradable characteristics of these metals, a subsequent surgery is typically necessary to extract the implant following the recovery or regeneration of the bone. Consequently, the demand for innovative biomaterials has surged in recent years, necessitating materials that are biodegradable within the physiological environment of the human body and do not require a secondary surgical procedure for implant removal during bone regeneration. Magnesium (Mg) and zinc (Zn) are considered optimal biodegradable materials due to their non-toxic absorption in the human body, prompting researchers to address the demand [3] [4]. However, studies on welding of Magnesium and zinc by conventional methods are not attempted by researchers due to the existing wide variation in properties, explosive welding offers a reliable solution. In this study an attempt is made to numerically simulate explosive welding of Mg and Zn sheets in parallel (Fig.1a) and inclined (Fig.1b) configuration and the variation in pressure, velocity, and strain are reported.

Numerical Simulation

A SPH (Smoothed Particle Hydrodynamics) method [5][6] using 2D AUTODYN (ANSYS 2020 R1) was developed to determine the magnesium-zinc weld behaviour, employed as flyer and base plates respectively. Zinc, which positioned as the base plate, has a density of 7.14 kg/m³, while magnesium, serves as the flyer plate, has a density of 1.74 kg/m³. The dimension of the base plate

Content from this work may be used under the terms of the Creative Commons Attribution 3.0 license. Any further distribution of this work must maintain attribution to the author(s) and the title of the work, journal citation and DOI. Published under license by Materials Research Forum LLC.

is 105 mm x 5 mm and the flyer plate has dimensions of 105 mm x 2 mm respectively. A parallel and inclined setups, as seen in Fig. 1, were attempted for the parallel and inclined configurations whose parametric conditions are presented in Table 1. To study the impact of initial angle (0° and 10 °), a uniform 4 mm PETN explosive (detonation velocity of 5170 m/s) thickness was maintained for the simulation.

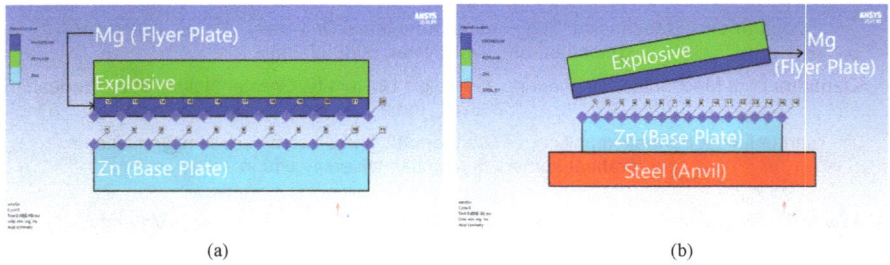

(a)   (b)

Fig. 1. 2D model configuration for (a) parallel setup (b) inclined setup

Table 1. Process Parameters

Combination	Test No	Explosive Thickness [mm]	Standoff distance [mm]	Inclination Degree [°]
Mg-Zn	1	4	3	0
Mg-Zn	2	4	3	10

The gauges are placed on the collision faces of the base plate and flyer plate to determine the values for pressure, Y-velocity and strain energy during the controlled energetic explosion. Particle size of 50 μm is employed for the SPH model acquiring 2,11,996 particles to illustrate the welding phenomenon.

Constitutive models and equations of state

The Jones-Wilkins-Less (JWL) [7] equation describes the explosive behaviour with respect to pressure, volumetric strain and internal energy. The governing equations is represented as Eq. 1

$$\ddot{p} = A\left(1 - \frac{\omega}{R_1 V}\right)e^{-R_1 V} + B\left(1 - \frac{\omega}{R_2 V}\right)e^{-R_2 V} + \frac{\omega E_s}{V} \qquad (1)$$

Where, p is the pressure, V denotes volume, E_s denotes internal energy, ω denotes the Gruneisen coefficient. R1, R2, A and B are constants, and is presented in the Table 2.

Table 2. JWL equation coefficients for explosives

A [GPa]	B [GPa]	R_1	R_2	ω
348	11.28	7	2	0.88

In establishing the relation between density (ρ), pressure (P), internal energy (E) and the pressure exerted on the flyer plate and base plate during compression the Mie-Gruneisen equation [8][9] is adopted and is represented by Eq. 2

$$P = \begin{cases} \dfrac{\rho_0 c_0 \mu \left[1+\left(1-\frac{\gamma}{2}\right)\mu-\frac{\alpha_1\mu^2}{2}\right]}{\left[1-(S_1-1)\mu-\frac{S_2\mu^2}{\mu+1}-\frac{S_3\mu^3}{(\mu+1)^2}\right]^2} + (\gamma+\alpha_1\mu)E_0; \ \mu > 0, \\ \rho_0 C_0^2 \mu + (\gamma+\alpha_1\mu)E_0; \qquad\qquad \mu \leq 0 \end{cases} \tag{2}$$

Where, ρ_0 is initial density of material; c_0 is sonic speed of the material; $\mu = \frac{\rho}{\rho_0} - 1$; $\gamma_0 =$ the Gruneisen coefficient, E_0 is the internal energy of the metal per unit mass and S_1, S_2, S_3 are the material constants as shown in Table 3.

Table 3. Mie-Gruneisen equation coefficient of the metals

Material	C_0 [m/s]	S_1	S_2	γ_0
Magnesium	5421	1.26	0	1.42
Zinc	4125	1.58	0	1.96
Steel	5431	1.49	0	2.17

The Johnson-Cook constitutive model [10][11] define the material behaviour of the flyer and base plates, alongside the stress and strain rate. The model is defined by Eq. 3

$$\alpha = \left(A + B\varepsilon_{eff}^n\right)(1 + C \ln \dot{\varepsilon})(1 - T^{*m}) \tag{3}$$

Where, $\alpha =$ flow stress; $\varepsilon_{eff} =$ effective plastic strain; A,B,C,n,m = material parameters; $\dot{\varepsilon} = \frac{\varepsilon_{eff}}{\dot{\varepsilon_0}}$, plastic strain rate; $T^* = \frac{T-T_{room}}{T_{melt}-T_{room}}$, homologous temperature. The values of the constants are presented in Table 4.

Table 4. Material constants

Materials	C	n	m
Magnesium	0.04	0.2	0.01
Zinc	0.01	0.3	0.05
Steel	0.076	0.643	0.5

Results and Discussion

Pressure

The pressure distribution at varied setup arrangement facilitates the bonding mechanism and the formation of a wavy structure at the interface. The pressure during the initial detonation phase is minimal; however, around 4.5×10^{-3} ms, the pressure escalates, attaining a maximum peak pressure of 3×10^4 MPa for the parallel configuration and 4×10^4 MPa for the inclined configuration. This increase in pressure is due to the kinetic energy imparted to the flyer plate during the high energetic explosion and helps in producing a wavy structure at the interface as seen in Fig. 2b. This phenomenon of gradual increase in pressure initially and reached the peak value at the mid region is similar to Ayele et al.[12].

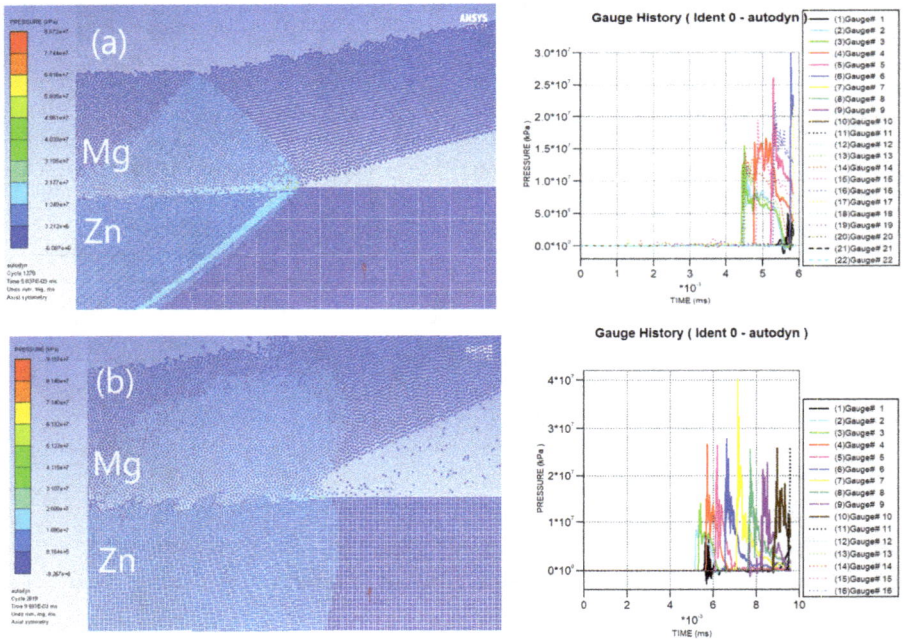

Fig. 2. 2D pressure contour and pressure v/s time graph of explosive height of 4 mm (a) parallel setup (b) inclined setup

Velocity Distribution

The flyer plate velocity plays a crucial role in affecting the bonding of the interface, appropriate magnitude of the flyer plate leads to deformation at the interface due to the high velocity impact causing jetting to occur welding. The flyer plate velocity[6] (721 m/s) obtained by analytical calculation (Eq. 4)

$$V_{PY} = V_d \sin \beta \tag{4}$$

Where dynamic bend can be calculated by Eq. 5

$$\beta = \frac{\pi}{2}\left(\sqrt{\frac{k+1}{k-1}} - 1\right)\left(\frac{R}{R+2.71+0.184\frac{t_e}{S}}\right) \tag{5}$$

Where, k is a parameter varying between 1.96 and 2.80, S is the standoff distance, and t_e is the explosive thickness.

The plate velocity is consistent to the maximum plate velocity 750 m/s obtained in numerical simulation conducted as seen in Fig. 3a and a maximum plate velocity around 1130 m/s as seen in

Materials Joining and Manufacturing Processes: MJMP 2025   Materials Research Forum LLC
Materials Research Proceedings 55 (2025) 27-33   https://doi.org/10.21741/9781644903612-5

Fig. 3b. The difference in the velocity distribution at the point of collision and the periodic disturbances in the materials enables in the formation of undulating interfaces at the bonding interface as reported by Yang et al. [13].

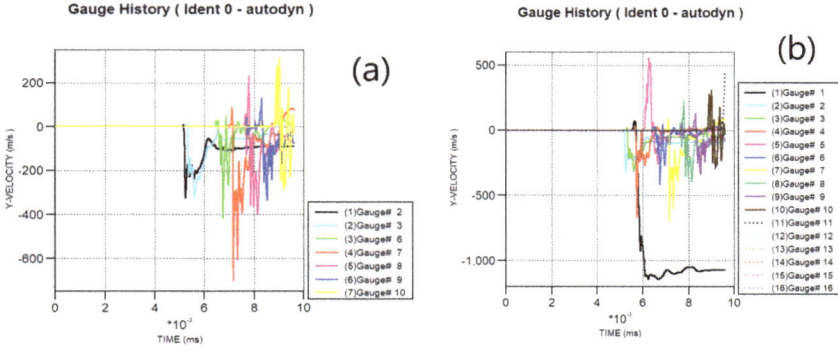

Fig. 3. Velocity distribution graph in Y-Velocity v/s time (a) parallel setup (b) inclined setup

Effective Plastic Strain

The effective plastic strain raises the plastic deformation over time[14]. This, high plastic deformation results in significant plastic strain due to the high pressure generated during detonation[15]. The bonding mechanism of explosive welding is primarily influenced by the jetting condition, which results in the robust plastic deformation at the interface depicted in Fig. 4.

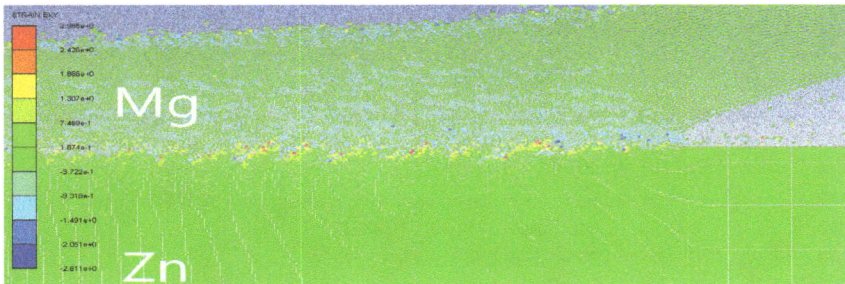

Fig. 4. Effective plastic strain along the interface

Waves are generated at the interface of the colliding metal plates as a result of a phenomenon known as "jet formation." This phenomenon involves the creation of a high-velocity jet of metal between the plates, which in turn causes indentations and protrusions that result in a wavy pattern as the jet interacts with the opposing surface[16]. Increasing the inclination angle, increases the dynamic bend angle which helps in attaining more high energetic impact on the flyer plate. This process is essentially a "relay race" of successive indentations on each plate. Jetting primarily originates from the flyer plate, attributed to its lower density compared to the denser base plate. The simulation results indicate a wavy interface resulting from the high-pressure shock wave exerted on the flyer plate, which induces atomic diffusion at the surface, resulting in permanent plastic deformation at the interface to achieve optimal bonding strength.

Conclusion

The simulation study conducted using ANSYS AUTODYN 2D for a parallel and inclined setup resulted successful bonding. The inclined setup exhibits waviness at the interface, whereas the parallel setup shows straightness. This variation at the interface results from the elevated pressure of (4×10^4 MPa) in the inclined configuration, in contrast to the pressure of (3×10^4 MPa) in the parallel configuration. The velocity distribution profile corelates to the flyer plate velocity required for jetting to occur at the interface enabling plastic deformation at the surface for bonding.

References

[1] F. Xing, S. Li, D. Yin, J. Xie, P.M. Rommens, Z. Xiang, M. Liu, U. Ritz, Recent progress in Mg-based alloys a a novel bioabsorbable biomaterials for orthopedic applications, J. Magn. and Alloys. 10 (2022) 1428-1456. https://doi.org/10.1016/j.jma.2022.02.013

[2] H. Kabir, K. Munir, C. Wen, Y. Li, Recent research and progress of biodegradable zinc alloys and composites for biomedical applications: Biomechanical and biocorrosion perspectives, Bio. Mater. 6 (2021) 836-879. https://doi.org/10.1016/j.bioactmat.2020.09.013

[3] S. Zhang, X. Zhang, C. Zhao, J. Li, Y. Song, C. Xie, H. Tao, Y. Zhang, Y. He, Y. Jiang, Y. Bian, Research on an Mg-Zn alloy as a degradable biomaterials, Acta Biomat. 6 (2010) 626-640. https://doi.org/10.1016/j.actbio.2009.06.028

[4] R. Krishnan, S. Pandiaraj, S. Muthusamy, H. Panchal, Md.S.Alsoufi, A.Md.M. Ibrahim, A. Elsheikh, Biodegradable magnesium metal matrix composite for biomedical implants: synthesis, mechanical performance, and corrosion behaviour-a review, J. Mater. Res. and Tech. 20 (2022) 650-670. https://doi.org/10.1016/j.jmrt.2022.06.178

[5] S. Saravanan, An experimental investigation on the explosive plugging of similar and dissimilar steel tubes, weld.Intern. 38(2) (2024) 128-139. https://doi.org/10.1080/09507116.2023.2293900

[6] P. Kumar, S. K. Ghosh, S. Saravanan, J.D. Barma, Experimental and simulation studies on explosive welding of AZ31B-Al 5052 alloy, Inter.J. adv.Man.Tech. 127 (2023) 2387-2399. https://doi.org/10.1007/s00170-023-11684-8

[7] C. Mortensen, P.C. Souers, Optimizing code calibration of the JWL explosive equation-of-state to the cylinder test, Prop. Expl. Pyro. 42(6) (2017) 616-622. https://doi.org/10.1002/prep.201700031

[8] Z.L Zhang, M.B Liu, Numerical studies on explosive welding with ANFO by using a density adaptive SPH method, J. Manuf. Proces. 41 (2019) 208-220. https://doi.org/10.1016/j.jmapro.2019.03.039

[9] C.W.DKumar, S. Saravanan, K. Raghukandan, Numerical and experimental investigation on aluminium 6061-V-grooved stainless steel 304 explosive cladding, Def. Tech. 18 (2022) 249-260. https://doi.org/10.1016/j.dt.2020.11.010

[10] J. Wang, X.J. Li, Y.X.Wang, Experimental and numerical study on the explosive welding of niobium-steel, Adv. Manuf. Tech. 122 (2022) 1857-1867. https://doi.org/10.1007/s00170-022-09984-6

[11] S. Saravanan, Thermo-structure approach to dissimilar explosive cladding with interlayer, J. Phy.: Conf. series. 2478(4) (2023) 042014. https://doi.org/10.1088/1742-6596/2478/4/042014

[12] B.S. Ayele, B.A. Mebratie, A.A. Meku, Investigation of the effect of explosive welding parameters in aluminium-steel bimetals: a numerical approach, J. Mater. in Civil Engi. 35 (2023) 04023415. https://doi.org/10.1061/JMCEE7.MTENG-16085

Materials Joining and Manufacturing Processes: MJMP 2025   Materials Research Forum LLC
Materials Research Proceedings 55 (2025) 27-33   https://doi.org/10.21741/9781644903612-5

[13] M. Yang, J. Xu, H. Ma, M. Lei, X. Ni, Z. Shen, B. Zhang, J. Tian, Microstructure development during explosive welding of metal foil: morphologies, mechanical behaviours and mechanism, Comp. Part B. 212(2021) 108685.
https://doi.org/10.1016/j.compositesb.2021.108685

[14] J.Feng, P. Chen, Q. Zhou, K. Dai, E. An, Y. Yuan, Numerical simulation of explosive welding using smoothed hydrodynamics methods, Int. J. of Multiphys. 11(3) (2017) 315-326.
https://doi.org/10.21152/1750-9548.11.3.315

[15] M, Fan, Effect of original layer thickness on the interface bonding and mechanical properties of Ti/Al laminate composites, Mater. Des. 99 (2016) 535-542.
https://doi.org/10.1016/j.matdes.2016.03.102

[16] I.A. Bataev, S. Tanaka, Q. Zhou, D.V. Lazurenko, A.M. Jorge Junior, A.A. Bataev, K. Hokamoto, A. Mori, P. Chen, Towards better understanding of explosive welding by combination of numerical simulation and experimental study, Mater. & Des.169 (2019) 107649.
https://doi.org/10.1016/j.matdes.2019.107649

Materials Joining and Manufacturing Processes: MJMP 2025   Materials Research Forum LLC
Materials Research Proceedings 55 (2025) 34-39   https://doi.org/10.21741/9781644903612-6

Evaluating equipment-dependent forging behavior of aluminum alloy

Akash Mahato[1,a], Saurabh Shrivastava[1,b], Sujit Goswami[1,c],
Suman Kumar Pandey[1,d], and Rahul Ramesh Kulkarni[2,e *]

[1](ADC-Forge Technology), Department of Foundry and Forge Technology, National Institute of
Advanced Manufacturing Technology [Formerly, National Institute of Foundry and Forge
Technology], Hatia, Ranchi-834003. Jharkhand. India

[2]Department of Foundry and Forge Technology, National Institute of Advanced Manufacturing
Technology [Formerly, National Institute of Foundry and Forge Technology], Hatia, Ranchi-
834003. Jharkhand. India

[a]Akamahato@gmail.com, [b]Saurabhicfai6@gmail.com, [c]sujeetgoswami6@gmail.com,
[d]sumanpandey.ranchi@gmail.com, [e]rahulkulkarni16122012@gmail.com

Keywords: AA6082, Forging Equipment, Cold and Warm Forging, Structure-Property
Correlation

Abstract. Aluminum Alloy 6082, a member of the Al-Si-Mg series, is widely recognized for its
excellent forgeability, making it a material of choice in critical applications such as the automotive
and aerospace industries. This alloy is utilized in both cast and wrought forms, offering versatility
in manufacturing processes. The present research investigates the influence of forging equipment
on the microstructural evolution and hardness properties of AA6082. Experiments were conducted
using different forging equipment, including a hydraulic press and a hammer, under both room
temperature and warm temperature conditions. The study highlights the dependency of
microstructure and hardness on the type of forging equipment employed, aiming to establish
correlations between equipment-induced variations and the alloy's mechanical performance. The
findings provide valuable insights into optimizing forging processes for AA6082, ensuring
enhanced performance in its intended applications.

Introduction
Aluminum alloy 6082 is the key material belonging to a group of Al-Si-Mg alloy whose having
better forgeability among many existing aluminum alloys. This alloy has medium strength due to
content of Mn as compared to AA6063/AA6061which result in better strength to weight ratio
having low density of aluminum as compared to other material such as Ti and Fe. These are having
applications in aerospace and automobile applications as a structural part such as suspensions
chassis etc.in cast as well as in wrought condition [1-6]. The AA6082 alloy has a melting point of
approximately 550 °C, and its hot forging is typically performed at a temperature range of 280–
300 °C. Considering this warm forging temperature is below 250 °C. These are heat treatable and
two-phase alloys with Mg_2Si as predominant second phase which affect the mechanical properties
of these alloys [7-11]. The forging is the conventional plastic deformation process used for
different products in industrial environment. There are different kind of forging process have been
used conventionally with different equipment such as hydraulic press, hammer etc. [12]. The
improvement of forgeability of the different alloys is possible by changing the microstructure
which reduces the cost of the component [13-14]. The previous study in reference to AA6082 [3,
15-16] indicate the solutioning associated with forging accompanying with other bulk metal
processes. There will be narrow range of study available to study the effect of equipment on cold
and warm forging behavior especially [17]. Leaping at this opportunity, the present investigation
aims at objectives is to study the effect of grain structure evolution due to cold and warm forging
of AA6082 under hydraulic press as well as hammer and to predict possible causes of grain

Content from this work may be used under the terms of the Creative Commons Attribution 3.0 license. Any further distribution of
this work must maintain attribution to the author(s) and the title of the work, journal citation and DOI. Published under license by Materials
Research Forum LLC.

Materials Joining and Manufacturing Processes: MJMP 2025 Materials Research Forum LLC
Materials Research Proceedings 55 (2025) 34-39 https://doi.org/10.21741/9781644903612-6

structure evolution and hardness values due to effect of these equipment with future scope for understanding the behavior of this materials in different forging condition.

Materials and methods

The material used for present research is AA6082 in T-6 heat treated condition whose chemical composition measured by weight chemical analysis. The specimen size was used 20 x 20 x20 in mm and deformed to 50% in open die forging process. The equipment used for open die forging are hydraulic press of 150 tonne and Pneumatic hammer (Manufacturer: FIELDING). The hardness was measured under Vickers hardness testing machine (Manufacturer: Fine Manufacturing Industries) with 5 kg load and 10 sec dwell time. The inverted metallurgical microscope (Make: Radical Scientific equipment) attached with Radical Metal 11.1 image analysis software was used. The muffle furnace with ± 2 °C variation was used to heat the material at 200 °C. For microstructure study, usual metallography technique such as dry polishing followed by wet polishing using alumina colloidal solution was used on the specimen and etched to reveal the microstructure with 5 ml HF, 10 ml H_2SO_4, and 85 ml water. The average grain size was calculated using line intercept method at 5 places of the specimens. The values of hardness and average grain size is reported with 95% CI within error bar of ±5%.

Results and discussion

Chemical Composition

The material used for investigation was AA6082 alloy in T6 heat treatment condition which is purchased from Bharat Aerospace Metals Mumbai. The chemical analysis of as received materials was done using weight chemical analysis. The result of chemical composition is given the table No. 1 which is given below:

Table 1
Chemical composition of AA6082-T6 aluminium alloy (wt%).

Mg	Si	Mn	Cu	Cr	Zn	Ti	Fe	Al
0.70	0.90	0.40	0.09	0.02	0.08	0.03	0.46	Balance

The above chemical composition is lie within the range of AA6082 as specified by ASTM standard. This confirms the alloy under investigation is AA6082.

Microstructure and hardness study on as received material AA6082

Fig. 1(a) shows the optical micrographs of as received material. The average grain size by line intercept method was found to be 42.04 μm. The micrograph reveals that there is variation in grain sizes. This can be validated by the plot (Fig. 1(b)) number of grain sizes verses grain size number. The grain size number by planimetric method varies from 5 to 14.5 with corresponding number of grains from 6 to 45. This variation causes the un-uniform mechanical properties. The average hardness was found to be 126.4968 in VHN.

This material further used to study the evolution of microstructure and hardness during different conditions of open die forging on the different equipment such as hydraulic press and hammer.

Materials Joining and Manufacturing Processes: MJMP 2025   Materials Research Forum LLC
Materials Research Proceedings 55 (2025) 34-39   https://doi.org/10.21741/9781644903612-6

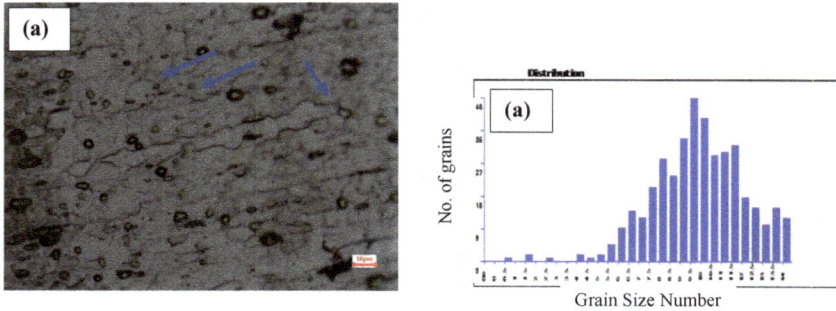

Fig. 1 *(a) Optical micrograph of as received AA6082 with (b) distribution plot of number of grains vs grain size number. (Condition A)*

Effect of temperature and equipment on AA6082 during open die forging

The as-received materials AA6082 further open die forged at room temperature and at warm temperature i.e. 200 °C on the hydraulic press and hammer to 50% deformation. The microstructure and hardness were evaluated on these conditioned and compared to understand the effect of equipment as well as temperature. The Fig. 2 shows the microstructure evolution during different conditions of open die forging. The condition A represented for as-received material, the condition B represented for cold forged material on hydraulic press, condition C represented for warmed forged material on hydraulic press, condition D represented for cold forged material on hammer and condition E represented for warm forge material on hammer. Investigations on these conditions imply the changes in the microstructure and distribution of grain size. The cold forged material shows the banded structure Fig. 2 (a), (b) and (a-1) (b-1) with bimodality of the grain structure. This result is more prominent in case of hammer forging at room temperature (Fig. 2 (b) and (b-1)). The reduction in grain size is observed in both the cases as compared to the as-received material (Fig. 3). This could be probably due to increase in mis-orientation angle during forging due to cold deformation. The grains revealed to be elongated with small grain size for the AA6082 cold forged under hammer. Interestingly, there is only slight variation in hardness probably due to distribution of second phases involved this kind of material rather than effect of grain size only which required further study for second phase.The as-received AA6082 warmed forged at 200 °C. The further decrease in grain size is observed during warm forging under hydraulic press as well as hammer. The decrease in grain size is more under hammer than that of under hydraulic press during warm forging of AA6082. In general, hydraulic press involve of gradual loading and hammer involve of sudden/impact loading. Due to this, in case of hydraulic press the time available for change of grain structure is more. Since, aluminium is high stacking fault energy material (166 mJ/m^2) as compared to other materials like magnesium, austenitic stainless steel, copper [18], recovery dominates during warm forging. So, the number of grains is more which are of different numbers in case of hydraulic press forging of AA6082. But, the un-uniformity in grain size observed to be in warm forged material under hydraulic press with more average grain size as compared to that of AA6082 warmed forged under hammer (Fig. 2(c), Fig. 2(c-1) and Fig. 3). This is not the case with hammer forging, since time available is very less for changing misorientation angle. The evolution of microstructure reveals the most uniformity in grain structure with decrease in bimodality among all the microstructure with average grain size of 16.1 μm ((Fig. 2(d), Fig. 2(d-1) and Fig. 3). Also, the hardness found to be less could be due to uniformity in grain size. But, for the understanding the hardness required study of second phase Mg_2Si as per previously studied literature [10-11].

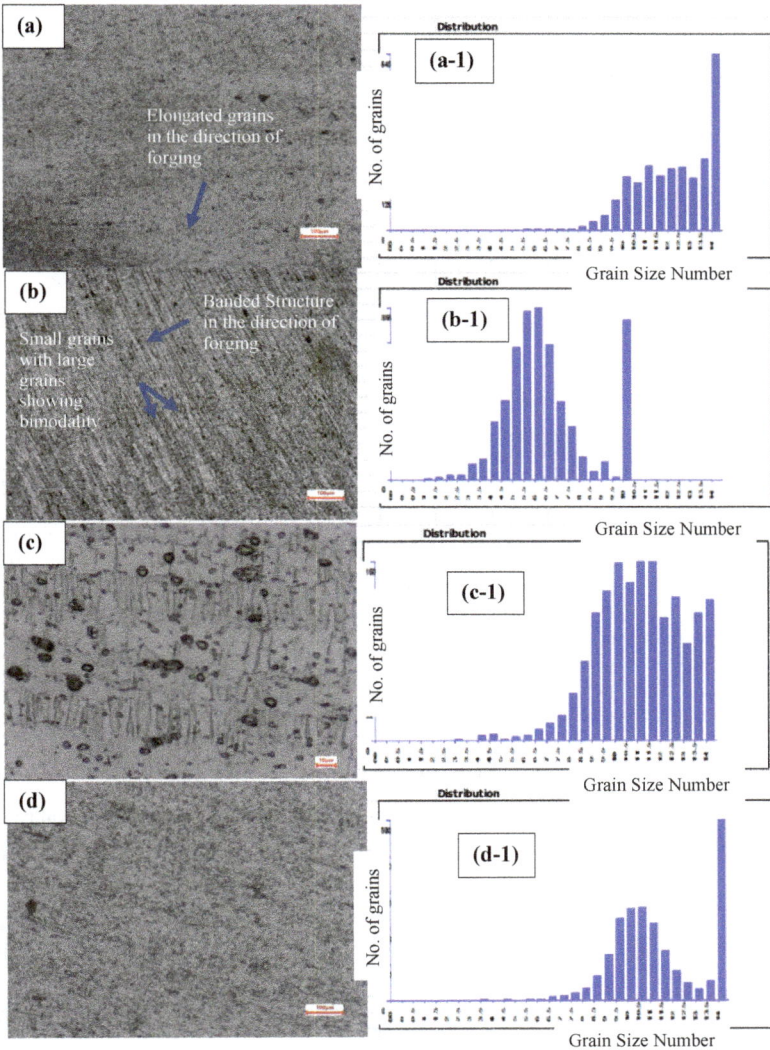

Fig. 2 *Optical micrographs evolved after (a) Cold forged AA6082 under hydraulic press (Condition B) with grain size distribution (a-1) (b) Cold forged AA6082 under hammer (Condition D) with grain size distribution (b-1) (c) Warm forged AA6082 under hydraulic press with grain size distribution (c-1) (Condition C) (d) Warm forged AA6082 under hammer with grain size distribution (d-1)(Condition E)*

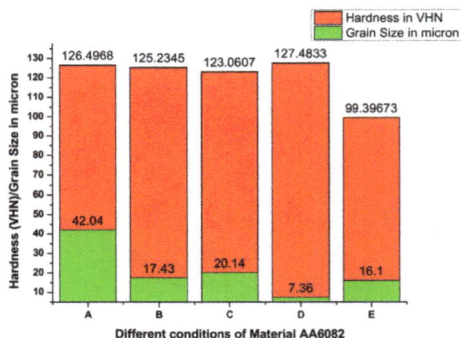

Fig. 3 *Bar chart showing hardness in VHN and grain size in micron at different condition of AA6082*

Conclusions

1. The unevenness of grain structure is observed in as received material which is in T6 heat treated condition.

2. The study concludes that forging conditions and equipment significantly impact microstructure evolution. Warm open-die forging at 200 °C with 50% deformation under a pneumatic hammer produces a uniform grain structure with an average size of 16.1 μm. Cold forging under the same conditions achieves the smallest average grain size of 7.36 μm, though with bimodal grain distribution, indicating improved forgeability. The pneumatic hammer demonstrates a stronger influence on grain structure evolution than the hydraulic press, aiding in selecting suitable equipment to optimize the mechanical properties of AA6082.There not much variation observed in hardness except condition of material under investigation i.e. for warm hammer forging. There is scope of further study of behaviour aspects of second phase as well as misorientation angles in grain structure at different conditions to understand the hardness values, observed during the current work.

Acknowledgement

The authors acknowledge the Government of India and the National Institute of Advanced Manufacturing Technology, Ranchi, for providing the necessary facilities to carry out this work.

References

[1] A. Heinz, A. Haszler, C. Keidel et al., Recent development in aluminium alloys for aerospace applications. Mater. Sci. Eng. A. 280(1) (2000) 102–107. https://doi.org/10.1016/S0921-5093(99)00674-7

[2] J. Hirsch, Recent development in aluminium for automotive applications. Trans. Nonferrous Metals Soc. China, 24(7) (2014) 1995–2002. https://doi.org/10.1016/S1003-6326(14)63305-7

[3] A. Ranjan, A. Kumar, R. Kulkarni, Studies on Annealing Kinetics of Cold Forged AA6082, In Dikshit MK, Soni A, Davim JP (eds) Advances in Manufacturing Engineering, Springer Nature, Singapore, 2022, 345-351. https://doi.org/10.1007/978-981-19-4208-2_24

[4] W. A. Monteiro, I. M. Espósito, R. B. Ferrari, S. J. Buso, Microstructural and mechanical characterization after thermomechanical treatments in 6063 aluminium alloy. Mater Sci Appl 2(11) (2011) 1529-1541. https://doi.org/10.4236/msa.2011.211206

[5] K. Strobel, M. A. Easton, L. Sweet, M. J. Couper, J. F. Nie. Relating quench sensitivity to microstructure in 6000 series aluminium alloys. Mater Trans 52(5) (2011) 914-919. https://doi.org/10.2320/matertrans.L-MZ201111

[6] J. R. Davis, Corrosion: Understanding the Basics. ASM International, 2000.

[7] A. R. Prabhukhot, K. Prasad, Effect of heat treatment on hardness of 6082-t6 aluminium alloy. Int J Sci Eng Res 6 (12) (2015) 38–42.

[8] A. R. Prabhukhot, effect of heat treatment on hardness and corrosion behaviour of 6082–t6 aluminium alloy in artificial sea water. Int J Mat Sci Eng 3(4) (2015) 287–294.

[9] E. Hennum, K. Marthinsen, U. H. Tundal, Effect of Microstructure on the Precipitation of β-Mg_2Si during Cooling after Homogenisation of Al-Mg-Si Alloys, Metals 2024, 14(2), 215. https://doi.org/10.3390/met14020215

[10] G. R. Ma, X. L. Li, L. Li, X. Wang, Q.F. Li, Modification of Mg_2Si morphology in Mg–9%Al–0.7%Si alloy by the SIMA process, Mat. Char. 62(3) (2011) 360-366. https://doi.org/10.1016/j.matchar.2011.01.006

[11] T. Ozkan, I. Tutuk, S. Acar, R. Gecu, K. A. Guler, Microstructural, hardness, and wear properties of Mg_2Si-reinforced Al matrix in-situ composites produced by low superheat casting, Mat. Today Commun.38 (2024) 107937. https://doi.org/10.1016/j.mtcomm.2023.107937

[12] J. Carvill, Manufacturing technology, Mechanical Engineer's Data Handbook (1993), 172-217.

[13] P. K. Ajeet Babu, M. R. Saraf, K. C. Voracious et al., Influence of forging parameters on the mechanical behaviour and hot forgeability of aluminium alloy. Mat. Today Proc. 2(4-5) (2015) 3238–3244. https://doi.org/10.1016/j.matpr.2015.07.132

[14] A. Kumar, R. R. Kulkarni, R. Ohdar, Towards understanding the behaviour of magnesium alloy during different forging processes: An overview (in press), Mat. Today Proc. (2023) DOI: https://doi.org/10.1016/j.matpr.2023.03.448

[15] N. Zhao, H. Ma, Z. Hu, Y. Yan, T. Chen, Microstructure and mechanical properties of Al-Mg-Si alloy during solution heat treatment and forging integrated forming process, Mat. Char. 185 (2022) 111762. DOI: https://doi.org/10.1016/j.jmatprotec.2022.117715

[16] N. Zhao , H. Ma , Q. Sun , Z. Hu , Y. Yan , T. Chen , L. Hua, Microstructural evolutions and mechanical properties of 6082 aluminium alloy part produced by a solution-forging integrated process, , J Mat. Proc. Tech.308 (2022) 117715. https://doi.org/10.1016/j.jmatprotec.2022.117715

[17] R. R. Kulkarni, V. L. Chakote, V. V. Shevale, Comparative Study on Microstructure Evolution during Cold Forging and Warm Forging in AA6082. NanoWorld J 9 (S1) (2023) S162-S165. DOI: 10.17756/nwj.2023-s1-033

[18] B. Hammer, M. C. Payne, Stacking fault energies in aluminium, J Phy.: Condensed Matter 4(50) (1999) 10453. https://doi.org/10.1088/0953-8984/4/50/033

Materials Joining and Manufacturing Processes: MJMP 2025   Materials Research Forum LLC
Materials Research Proceedings 55 (2025) 40-44   https://doi.org/10.21741/9781644903612-7

Modification of dye/fluorescent penetrant testing in accordance with Industry 4.0

Nitish Kumar[1,a] *, Banshidhara Mallik[2,b] * and Rahul Ramesh Kulkarni[3,c]

[1]Department of Materials and Metallurgical Engineering, National Institute of Advanced Manufacturing Technology [Formerly, National Institute of Foundry and Forge Technology], Hatia, Ranchi-834003. Jharkhand. India

[2]Department of Materials and Metallurgical Engineering, National Institute of Advanced Manufacturing Technology [Formerly, National Institute of Foundry and Forge Technology], Hatia, Ranchi-834003. Jharkhand. India

[3]Department of Foundry and Forge Technology, National Institute of Advanced Manufacturing Technology [Formerly, National Institute of Foundry and Forge Technology], Hatia, Ranchi-834003. Jharkhand. India

[a]bansidharamallik@gmail.com, [b]bansidharamallik@gmail.com, [c]rahulkulkarni16122012@gmail.com

Keywords: Dye/Florescent Techniques, Discontinuity, Automation in Dye Penetrant Test, Industry 4.0

Abstract. Dye/liquid fluorescent penetrant testing is a non-destructive testing method. It is used to detect discontinuities exposed to the surface in engineering components and metals involving different manufacturing processes. This approach depends on the physical interplay between a tailored chemical liquid and the surface of the component being tested. As a result of this interaction, the liquid penetrates surface cavities and then emerges, providing a visual indication of the location and approximate dimensions of the openings. Several steps are involved in conducting this test. As a final procedure visual/mechanized (through different magnifying systems) is involved to detect the defects/flaws. The defects like laps, porosity, cracks, seams and other surface discontinuities can be detected speedily with high degree of reliability. Mechanization and automation of viewing the process is one of the aspects of Industry 4.0. This can be achieved with suitable devises to monitor the behavior of penetrant on the surface with the help of automatic cameras available in present scenario. This article reviews the automation aspects involved in dye penetrant test with modern techniques.

Introduction

Dye or fluorescent penetrant testing (DPT) is a non-destructive evaluation technique applied to detect with reveal of flaws on surface in various materials used for engineering applications and finished products, aiding in the elimination of defective parts. The method works by utilizing a specially formulated chemical liquid that interacts physically with the part's surface, seeping into cavities on the surface and resurfacing to reveal visually the size and location of the openings. The objective of dye penetrant testing/FPI (Fluorescent Penetrant Inspection) is to facilitate rapid and reliable visual evidence of surface defects such as cracks, porosity, laps, seams, and other flaws. The real challenge in traditional method is time taken to perform the test manually (1 hr. for small test piece consumed approximately). This can be minimized with modification and automation.

With the advent of Industry 4.0, advanced technologies like automation, artificial intelligence, and robotics are bridging the physical and digital realms of manufacturing [1-4]. This paper proposes incorporating Industry 4.0 technologies into DPT by introducing automated data recording systems, including cameras, videography, and interconnected sensors for efficient data

Content from this work may be used under the terms of the Creative Commons Attribution 3.0 license. Any further distribution of this work must maintain attribution to the author(s) and the title of the work, journal citation and DOI. Published under license by Materials Research Forum LLC.

Materials Joining and Manufacturing Processes: MJMP 2025   Materials Research Forum LLC
Materials Research Proceedings 55 (2025) 40-44   https://doi.org/10.21741/9781644903612-7

storage [1-4]. These systems enable seamless data collection, exchange, and analysis, potentially enhancing productivity and manufacturing efficiency.

This article discusses with aim considering 1. There is gap in between normal testing for defect visualization so far as Industry 4.0 is concerned, which need the speedy data generation and recording for future evaluations. 2. To modify the present evaluation method by incorporating the data generation and recording system which is proposed.

Procedural steps in DPT and their modifications relevance to Industry 4.0
Conventionally, this test consists of conventional eight steps [5-7] shown in following flow chart to be carryout on the parts to be inspected.

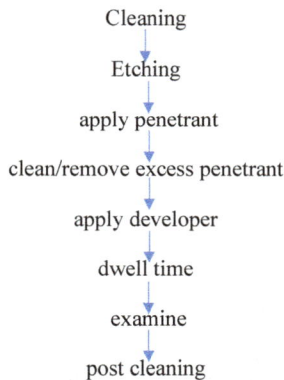

Cleaning

↓

Etching

↓

apply penetrant

↓

clean/remove excess penetrant

↓

apply developer

↓

dwell time

↓

examine

↓

post cleaning

Flow chart showing traditional steps involved in DPT/ FPI

1. Pre-cleaning and drying the test surfaces of the object under examination
2. Application of liquid penetrant to the test surfaces uniformly and permit it to dwell on the part surface for period of time (may vary from metal to metal) to allow it enter and making sure that, any discontinuity fill completely
3. Excess liquid penetrant is to be removed from the test surfaces by suitable means (cottons/ by using different the suitable penetrant such as water washable, solvent removable, emulsifier etc.)
4. The application of a developer serves to extract the held penetrant from discontinuities, spreading it on the test surface to improve flaw detection. It intensifies the brightness of indications and provides a contrasting surface for visible dye indications during fluorescent testing. Then, visual examination of surfaces for liquid penetrant indications, interpretation is to be carried out.
5. Post cleaning of the part for any unwanted materials

The above conventional process steps have been extended/replaced for automation in Industry 4.0 especially steps number 5 which has been discussed further. In this, data acquisition and recording (like videography of the process) can be involved so far as this testing is concerned.

The modified steps with in accordance with Industry 4.0 is shown in following flow chart

Materials Joining and Manufacturing Processes: MJMP 2025   Materials Research Forum LLC
Materials Research Proceedings 55 (2025) 40-44   https://doi.org/10.21741/9781644903612-7

Cleaning

Etching

apply penetrant

clean/remove excess penetrant

apply developer

dwell time

Automated Pictures and Automated evaluation

Recoding and storage with expert validation if required

post cleaning

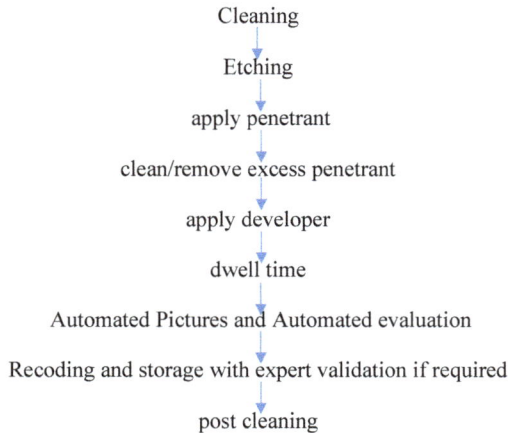

Flow chart chowing modified steps involved in DPT/ FPI in reference to Industry 4.0

Mechanized Scanning of indications raised during FPI

Mechanized scanning of fluorescent liquid penetrant test indications aims to automate the visual inspection and analysis components of the test procedure. In this connection multi robot collective system software developed by J. Karigiannis et al. [8]. In this framework, the "inspecting agent" captures images at different stages of the automated FPI inspection, which are analyzed using a customized deep neural network model to classify indications. The multi-robot system 203 tests was conducted to generate the dataset on the multi-robot system for turbine engine components under actual manufacturing conditions to evaluate its performance.

Mechanized scanning systems employ robotic arms or automated platforms to systematically move an imaging device over the component's surface. Key components of such systems include:

- UV Light Sources: Enhance visibility of fluorescent indications.
- Cameras with better resolution: Capture detailed images of the surface.
- Image Processing Algorithms: Analyze the captured images to detect and classify indications involve machine learning and deep learning
- Control Systems: Coordinate the motion and inspection parameters of the scanner.

These systems ensure comprehensive coverage and consistent inspection quality, even for complex geometries. Despite its advantages, mechanized scanning systems face challenges such as:

- High Initial Costs: Investment in equipment and setup can be substantial.
- Surface Complexity: Inspecting irregular or intricate surfaces requires advanced programming and equipment.
- Lighting Variability: Ensuring consistent UV illumination across varying inspection environments.

Ongoing research aims to address these challenges through innovations in adaptive scanning and AI-driven analysis.

Materials Joining and Manufacturing Processes: MJMP 2025
Materials Research Proceedings 55 (2025) 40-44

Materials Research Forum LLC
https://doi.org/10.21741/9781644903612-7

Machine Learning in FPT (Fluorescent Penetrant test)

Artificial Intelligence is one of the emerging filed in todays' scenario. There has been attempt made to apply the branches of artificial intelligence such as machine learning and deep learning in the FPT to detect the flaws for automation in Industry 4.0. One of the machine learning algorithms i.e. random forest proves that the flaws can be detected and distinguished with these algorithms with 76% accuracy as compared to human inspection in aerospace component [9-10]. But, for the further improvement, there is requirement of detailed study with more trained data sets for application in real life components.

In addition to random forest algorithm, neural network can be applied in the FPI. Such a work has been studied previously [9] for spot detection and image processing in FPI. This work uses the direct spot classifier as well as indirect spot classifier in which it claims that the former require for extractive features like linear defects, cracks etc. with existing trained data.

A similar study has been done [8] by applying multi robot collective system consisting of robotic arm, turn table and other manipulators which can be helpful in automation of FPI process. This way, it can modify the traditional FPI process commensurate as per Industry 4.0.

This robotic system mimics human inspection behavior, performing FPI under both UV and white light using brushing or wiping techniques to highlight indications. Multi-robot systems are anticipated to boost accuracy to 95% from 92% [8] and sensitivity to 96% from 92% [7]. Proof-of-concept trials in industries such as aviation could see these systems operating in shadow mode with human inspectors. Future initiatives will include additional experiments in manufacturing environments to validate and optimize the system's performance.

Summary

A novel process for DPT/FPI modifications is discussed in present article.

1. The main purpose of modified DPT/FPI is mechanized scanning of indications and to automate the viewing of the process. Moreover, the entire process, from sample cleaning to the final detection of flaws, can be automated and recorded, ensuring reliable and reproducible results. This allows all accessible surfaces to be examined in a shorter time even for complex geometrical parts.
2. In all these contexts, the use of artificial intelligence with machine learning and deep learning plays a very important role in todays' era.
3. There lies a challenge in detecting the surface flaws/discontinuities in porous materials with automation. The porous materials always a limitation in this test in general due to absorption of the penetrants. This can be visualized using proper data acquisition system based on size and shape of porosity.
4. AI, ML, and computer vision-based techniques can be utilized for future research direction to minimize the discrepancies as well as to improve the speed of the technique

References

[1] B. Liu et al., Study on the automatic recognition of hidden defects based on Hilbert Huang transfom and hybrid SVM-PSO model, Prognostics and System Health Management Conference, China (2017) 1-7. https://doi.org/10.1109/PHM.2017.8079294

[2] L. N. Smith, Cyclical Learning Rates for Training Neural Networks, IEEE winter conference on applications of computer vision (2017) 464-472. DOI: https://doi.org/10.48550/arXiv.1506.01186

[3] K. He et al., Deep Residual Learning for Image Recognition, IEEE Conference on Computer Vision and Pattern Recognition (CVPR), Las Vegas, NV, USA, 2016, pp. 770-778. https://doi.org/10.1109/CVPR.2016.90

[4] S. A. M. Hashim Et al., Automatic Weld Based Discontinuities Detection based on Dye Penetrant Test by Deep Learning, MyJICT (2025) 1-9. DOI: https://doi.org/10.53840/myjict10-1-141

[5] ASNT. *Non-Destructive Testing Handbook: Liquid Penetrant Testing.* American Society for Non-destructive Testing, 2018.

[6] ASTM E1417/E1417M-21. "Standard Practice for Liquid Penetrant Testing." ASTM International, 2021.

[7] G. Kedarnath, KVS Phani, R. K. Sahu, B. Kantharaju, Fatigue surface crack detection by using fluorescent dye penetrant test technique on welded engineering service components, Int. Res. J Eng. Tech. 4(4) (2021) 2103-2707.

[8] J. Karigiannis et al., Multi-robot System of Automated Florescent Penetrant Indication Inspection with Deep Neral Nets, Proc. Manufact. 53 (2021) 735-740. DOI: 10.1016/j.promfg.2021.06.072

[9] N. J. Shipway et al., Automated defect detection for fluorescent penetrant inspection using random forest, NDT and E int., 101 (2019) 113-123. DOI: https://doi.org/10.1016/j.ndteint.2020.102400

[10]   A. Niccolai et al., Machine Learning based detection technique for NDT in Industrial Manufacturing, Mathematics 9 (2021) 1251-1266. https://doi.org/10.3390/math9111251

Materials Joining and Manufacturing Processes: MJMP 2025   Materials Research Forum LLC
Materials Research Proceedings 55 (2025) 45-50   https://doi.org/10.21741/9781644903612-8

Mechanical performance of glass fiber epoxy laminates with embedded circular and square cutouts

Aravind MUDDEBIHAL[1,2,3,a], P.S. Shivakumar Gouda[4,5,b] *, Vinayak S. UPPIN[5,c], I. Sridhar[6,d]

[1]Research Center, Department of Mechanical Engineering, SDM College of Engineering &Technology, Dharwad, Visvesvaraya Technological University, Belagavi, Karnataka, India

[2]Department of Mechanical Engineering, Angadi Institute of Technology & Management, Belagavi, Visvesvaraya Technological University, Belagavi, Karnataka, India

[3]Junior Research Fellow, AR&DB, DRDO Research Lab, Department of Mechanical Engineering, SDM College of Engineering &Technology, Dharwad, Karnataka, India

[4]Department of Robotics and Artificial Intelligence, Mangalore Institute of Technology and Engineering, Moodbidri, Mangalore, Visvesvaraya Technological University, Belagavi, Karnataka, India

[5]AR&DB, DRDO Research Lab, Department of Mechanical Engineering, SDM College of Engineering &Technology, Dharwad, Karnataka, India

[8]Department of Mechanical Engineering, SDM College of Engineering &Technology, Dharwad, Visvesvaraya Technological University, Belagavi, Karnataka, India

[a]aravindbm21@gmail.com, [b]ursshivu@gmail.com, [c]ursuppin@gmail.com, [d]sridhari74@gmail.com

Keywords: Embedded Cutouts, Composite Laminates, Tensile, Buckling and FE Analysis

Abstract. This study explores the influence of induced embedded cutouts of circular and square shapes on tensile and buckling performance of glass fiber reinforced polymer (GFRP) composite laminates. Cutouts are often introduced in composite structures during manufacturing for functional requirements such as fastening, wiring, or integration with other components. But their presence creates a localized stress concentration that can significantly affect mechanical performance and failure behavior. Specimens with embedded circular, square shaped cutouts, and plain GFRP composite laminates were manufactured and tested under tensile and buckling load conditions to assess the extent of strength reduction. The results demonstrate that embedded circular holes cause a decrease in their tensile and buckling strengths up to 51% compared to plain laminates, as their geometry facilitates uniform stress distribution and reduces the likelihood of premature failure. In contrast, square cutouts lead to a more pronounced reduction up to 55% in both tensile and buckling strength, this was primarily due to the sharp corners serving as stress risers hence initiating damage and propagate the cracks. This study provides detailed insights into the behavior of GFRP laminates with embedded features and highlights the importance of optimizing the design for placement of cutouts through experimental and finite element analysis to balance functional requirements with mechanical reliability in engineering applications.

1. Introduction

Composite materials play a crucial rule in the aerospace and automotive sectors, by reducing weight, while simultaneously ensuring the strength, and enhanced performance. Cut-outs in composite structures features weight reduction and functionality but act as stress concentrators, risking crack initiation and failure. Proper placement of cutouts is crucial to maintain structural

Content from this work may be used under the terms of the Creative Commons Attribution 3.0 license. Any further distribution of this work must maintain attribution to the author(s) and the title of the work, journal citation and DOI. Published under license by Materials Research Forum LLC.

Materials Joining and Manufacturing Processes: MJMP 2025   Materials Research Forum LLC
Materials Research Proceedings 55 (2025) 45-50   https://doi.org/10.21741/9781644903612-8

integrity. Prior investigations have significantly examined the mechanical behavior of composite structures with cut-outs, focusing on their impact on load-bearing capacity and stress distribution. Among various shapes of cut-out, circular cut out verge to result in minimal stress concentration factor in both experimental and finite element simulation results. This makes circular cutout as a most preferred option in applications where minimizing stress risers is critical for maintaining structural integrity [1]. Lal et al. [2] studied stress concentration factor in an isotropic rectangular plate with a circular hole under a uniform tensile load through experimental and numerical approach. Effect of various cut outs shapes on stress concentration factor was studied on Steel plates. Circular cutouts were shown low stress concentration among all cutouts [3]. The high stress concentration in square and hexagonal cutouts was noticed due to presence of sharp corners [4]. Pan et al. [5] conducted experimental investigation on stress distribution in a finite plate containing a rectangular hole and applied Muskhelishvili's complex variable approach to validate their findings from finite element simulations. Mehmet [6] reviewed comparison study on lateral buckling strength of beams with and without cutouts, square cutouts showed better resistance over circular cutouts. Kumar et al. [7] reviewed effects of stress concentration in composite laminates due to cutouts, emphasizing on stress concentration factor from experiments and finite element approach. Further scope of research was identified as study on edge interactions, elliptical holes, and stiffeners on buckling strength. Gupta et al. [8] reviewed on the impact of cutouts on carbon fiber reinforced polymer composite laminates and reported reduced strength, delamination, and stress concentration. Hole size and laminate thickness influence damage and strength, with delamination either weakening or redistributing stress.

Previous studies showed that it is ideal to make circular cutouts on composite structures which results in low stress concentration and maintain structural integrity. Focusing on cutout shapes, orientation, sharp corners, and stress distribution studies revealed failure mechanism of composite laminates under lateral buckling loads. However, under influence of tensile and axial buckling loads, further investigation is required to highlight the effects of cutouts which is embedded during manufacturing of GFRP composite laminates.

2. Materials

2.1 Preparation of FRP laminate with embedded circular and square cutouts.

A 220 gsm UD glass fiber was cut to the required dimensions and placed onto a flat surface measuring 300 x 300 mm. To facilitate the convenient removal of tool plates from the moulds, a release agent was applied to prevent the fibers from bonding to the surface. Next, the liquid polymer, consisting of Epoxy Lapox L12 and Hardener K6, was mixed in the proportion of 10:1. The mechanical properties of E-glass and epoxy resin as specified by the supplier are listed in Table 1. The 55 wt. % of fibers were maintained for all the composite laminates.

Table 1: The material properties of the E-glass and Epoxy resin.

Mechanical Properties	E-glass	Epoxy Resin
Tensile strength (MPa)	2000	70-80
Modulus of elasticity (GPa)	72	4-4.8
Density (kg/m^3)	2500	1162

2.2 Sample preparation

Two wooden plies, each measuring 3x2 square feet, were chosen as the mould materials, as illustrated in Figure 1 (a) and (b). Metallic pins were employed to make a hole in the composite laminates during manufacturing. A hole has been drilled on top and bottom tool plates to create the cutouts in laminates. To avoid adhesion of matrix material to wooden moulds and inserts a

Materials Joining and Manufacturing Processes: MJMP 2025   Materials Research Forum LLC
Materials Research Proceedings 55 (2025) 45-50   https://doi.org/10.21741/9781644903612-8

releasing agent was applied. Totally 8 layers were stacked by spreading resin uniformly on the fiber mat and rolled properly to remove the entrapped air. While positioning the fiber mat, certain fibers were manually adjusted near the inserts to aid in forming the moulded hole, as depicted in Figure 1 (c). Further, the upper half of the mould was closed, and 50 kg of dead weight was placed on it for proper compaction. Further laminates were cured under room temperature for about 24 hours. Once the curing was complete, the composite laminate was taken out from the mould, as shown in Figure 1 (d). Similar procedure was followed to manufacture composite laminate with square cutout by inserting square shaped metallic pins. The specimen codes and cutouts dimensions are listed in Table 2.

Table 2. Specimen codes and cutout dimensions.

Specimens with Cirular cutout	Diameter (in mm)	Area of cutout (in mm^2)	Specimens with Square cutout	Side (in mm)	Area of cutout (in mm^2)
C1	3.0	7.0	S1	2.5	6.25
C2	5.5	24	S2	5.0	25.0
PS -Plane Sample	-	-	-	-	-

2.3 Testing

The tensile and buckling tests were performed as per ASTM D3039 [9] and ASTM E-2954-15 [10]. The universal testing machine (UTM) model of JE-AFUTM-10 kN/100 kN was used to conduct the tests. For each composition five samples were tested, and mean values were reported. All tests were carried by maintaining constant crosshead speed of 1 mm/min. For both tests, sample size of 250mm x 25mm x 2mm corresponding to its length, width, and thickness was followed respectively. To ensure a secure grip during tensile test, end tabs each 50 mm in length, were adhered to both ends of the specimen [11-12]. Further, sample loaded in UTM and testing was continued until failure of the specimen, and it can be observed as shown in Figure 2.

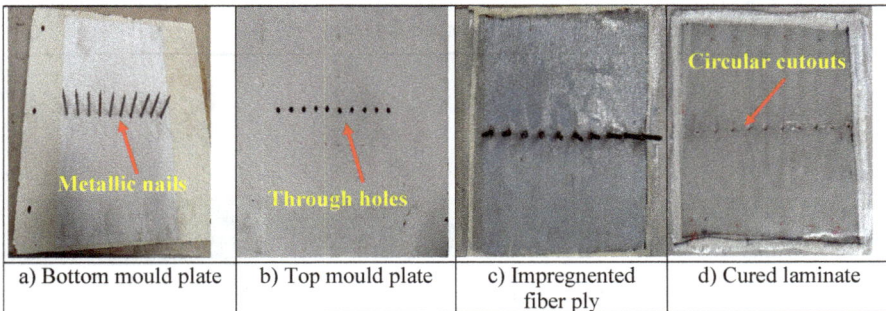

a) Bottom mould plate	b) Top mould plate	c) Impregnented fiber ply	d) Cured laminate

Figure 1: Embedded circular cutouts in GFRP composite laminate

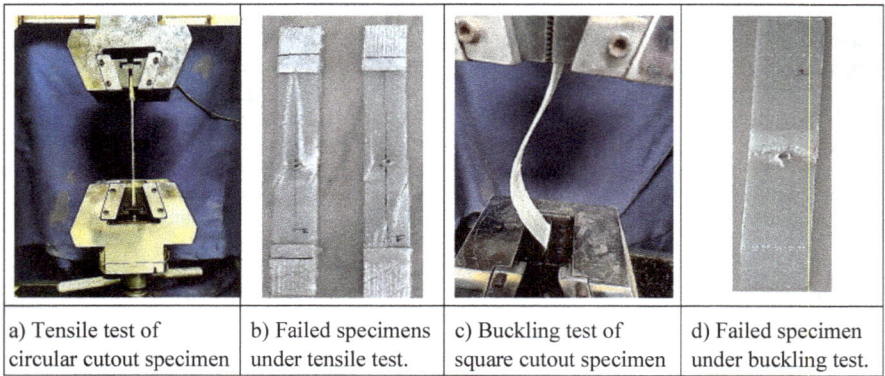

| a) Tensile test of circular cutout specimen | b) Failed specimens under tensile test. | c) Buckling test of square cutout specimen | d) Failed specimen under buckling test. |

Figure 2: Tensile and Buckling test of GFRP composite specimens.

2.4 FE Simulation

ABAQUS software was used to perform FE simulations. The boundary condition at lower grip, which is constrained, while a displacement load was applied to reference point of the upper grip to induce motion. A sufficiently refined mesh was employed to accurately capture the specimen behavior. The FE simulation utilized material properties derived from the experimental test results. The experimental data were validated against simulation results.

3. Results and Discussion

This section outlines the findings from experimental and simulation evaluation of GFRP composite specimens with circular and square cutouts, tested under tensile and buckling conditions. Figure 3 provides a visual representation of tensile test simulation result for the specimen C2 with a maximum stress of 179 MPa and from Figure 4 (a), the experimental findings are around 200MPa. The error between FEA and experimental test results showing approximately 10%, this may be due to errors in manufacturing and experimental testing. Similarly test results were recorded for all the specimens including buckling test and plotted in Figure 4 (b).

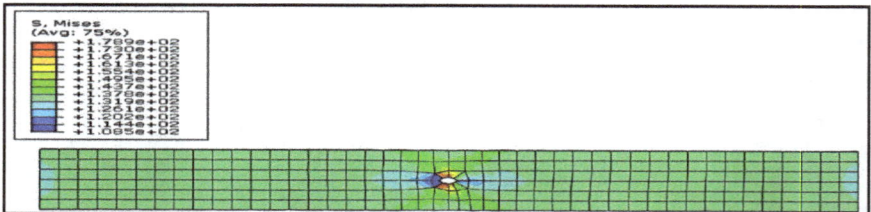

Figure 3: Tensile test simulation of specimen C2

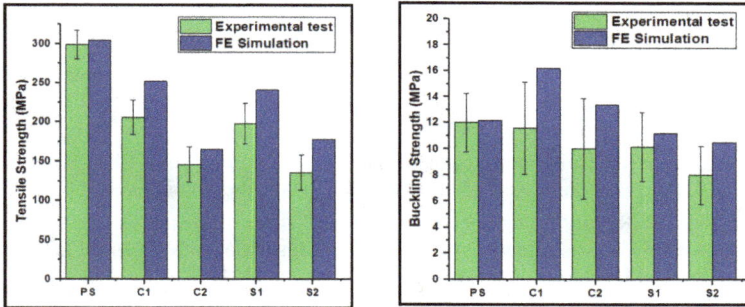

Figure 4: Strength of GFRP composite specimens under *(a)* Tensile *(b)* Buckling conditions

The effect of cutouts on tensile and buckling strength of GFRP composites as compared to the plain specimen (PS) and shown in Figure 4. A reduction in tensile and buckling strength was observed with an increase in the cutout area. Also, reduced tensile and buckling strength was observed in square cutouts compared to circular cutouts for the approximately same cutout area. This behavior was in strong correlation with FE simulation outcomes. Moreover, for comparing the same area of different cutouts, square cutout showed maximum drop about 10% and 13% under tensile and buckling loads respectively. This drop was attributed to the rise in stress concentration at the corners and sudden catastrophic failure.

4. Conclusions

This work provides essential information about the composite laminate-making procedure, considering both with and without cutouts during manufacturing.

- The presence of a cutout weakens the material's ability to withstand applied loads.
- Circular cutouts distribute stress more uniformly, reducing stress concentration and resulting in a strength reduction of up to 51%,
- whereas square cutouts, with sharp corners acting as stress risers, lead to higher localized stresses and reduced structural integrity, causing a strength reduction of up to 55%.
- Through experimental and FE studies it was optimized that the circular cutouts are preferred over square cutouts for the same cutout area.

Acknowledgement

The Authors would like to express their sincere gratitude to the organization for providing essential facilities at Research Center and AR&DB DRDO Research Lab, Department of Mechanical Engineering, SDM College of Engineering and Technology, Dharwad.

References

[1] S. D. Watsar, A. Bharule, Stress analysis of finite plate with special shaped cutout, International J. Sci. Engg. and Research, 3.4 (2015) 145-150. https://www.doi.org/10.70729/IJSER15116

[2] A. Lal, B. M. Sutaria, Rahul Kumar, Stress analysis of composite plate with cutout of various shape, IOP Conference Series: Materials Science and Engineering 814 (2020) 012011. 10.1088/1757-899X/814/1/012011

[3] J. Woo, W.B. Na, Effect of cutout orientation on stress concentration of perforated plates with various cutouts and bluntness, Intl. J. Ocean System Engg., 1.2 (2011) 95-101. https://doi.org/10.5574/IJOSE.2011.1.2.095

Materials Joining and Manufacturing Processes: MJMP 2025   Materials Research Forum LLC
Materials Research Proceedings 55 (2025) 45-50   https://doi.org/10.21741/9781644903612-8

[4] J. Rezaeepazhand, M. Jafari, Stress analysis of perforated composite plates, J. Composite Structures, 71 (2005) 463-468. https://doi.org/10.1016/j.compstruct.2005.09.017

[5] Z. Pan, Y Cheng, J. Liu, Stress analysis of a finite plate with a rectangular hole subjected to uniaxial tension using modified stress functions, Intl. J. of Mechanical Sci., 75 (2013) 265-277. https://doi.org/10.1016/j.ijmecsci.2013.06.014

[6] M. Bulut, Effectiveness of double cutout on lateral buckling analysis of fiber reinforced composites, J. Polymer Composites, 45 (2024) 10880-10887. https://doi.org/10.1002/pc.28515

[7] S. A. Kumar, R. Rajesh, S. Pugazhendhi, A review of stress concentration studies on fibre composite panels with holes/cutouts, Proceedings of the Institution of Mechanical Engineers, Part L: J. Materials: Design and Applications, 234.11 (2020) 1461-1472. https://doi.org/10.1177/1464420720944571

[8] S. Gupta, S. Pal, B. C. Ray, An overview of mechanical properties and failure mechanism of FRP laminates with hole/cutout, J. Applied Polymer Sci., 140.20 (2023) e53862. https://doi.org/10.1002/app.53862

[9] ASTM D3039, Test method for tensile properties of polymer matrix composite materials, ASTM International, West Conshohocken, PA, USA (2014).

[10] ASTM E2954-15, Standard Test Method for Axial Compression Test of Reinforced Plastic and Polymer Matrix Composite Vertical Members, 2022.

[11] M.J.M. Fikry, S. Ogihara, V. Vinogradov, The effect of matrix cracking on mechanical properties in FRP laminates, J.Mechanics of Advanced Materials and Modern Processes, 4 (2018) 1-16. https://doi.org/10.1186/s40759-018-0036-6

[12] I.M. Alarifi, Investigation into the structural, chemical and high mechanical reforms in B4C with graphene composite material substitution for potential shielding frame applications, J.Molecules, 26 (2021) 1921. https://doi.org/10.3390/molecules26071921.

Materials Joining and Manufacturing Processes: MJMP 2025   Materials Research Forum LLC
Materials Research Proceedings 55 (2025) 51-56   https://doi.org/10.21741/9781644903612-9

Optimization of process parameters in explosive welding using machine learning

Kusammanavar Basavaraj[1,a*], Satyanarayan[2,b], Anand Kulkarni[3,c]

[1]Department of CSE(AIML), Rao Bahadur Y Mahabaleswarappa Engineering College Ballari, Karnataka, 583104, India

[2]Department of Mechanical Engineering, Alva's Institute of Engineering and Technology Karnataka, Moodbidri, Mangalore 574225, India

[3]Department of Mechanical Engineering, Cambridge institute of Technology Technology, Karnataka, Bangalore, 560036, India

[a]basavaraj.k@rymec.in, [b]satyan.nitk@gmail.com, [c]anandkulkarni.mech@cambridge.edu.in

Keywords: Optimization, Explosive Welding, Machine Learning, Wavy Interface, Techniques

Abstract: A solid-state welding technique that joins two pieces of metal by controlled explosive detonation is called explosive welding (EXW), which has become a promising area of the study. However, it is well known that explosive welding is an expensive experiment. It is tough to expect the experimental results based on a practical approach by repeated attempts which are continued until success. In the present paper, though several Artificial Intelligence (AI) algorithms are implemented and trained using the dataset, the current state of AI algorithms based on the previous studies and their findings applied to the optimization of the welding process is reviewed and explained. Also, the types of optimization techniques available in order to predict the best results and most relevant input factors of explosive welding are reviewed. Based on the survey, the best optimisation technique is suggested for researchers.

Introduction

Researchers are looking forward to creating new materials to meet industrial standards. Solid-state welding techniques such as friction stir welding (FSW), explosive welding (EXW), diffusion welding, pressure welding etc., are used to combine similar or dissimilar metals/materials. Among these, the explosive welding (or detonation welding) technique is increasingly popular for joining materials/metals that are difficult/impossible to weld using traditional methods and other solid-state welding methods [1-4]. EXW technique joins two materials, mechanically and metallurgically by harnessing the energy of high-speed collisions. To establish a mechanical/metallurgical connection between the materials, the EXW process includes melting, plastic deformation, and atomic diffusion [5–7]. However, the process of generating data by experimenting is expensive. It is very difficult to predict the results based on a practical approach on a trial-and-error basis. Therefore, engineers need to generate an optimization method/technique that an experimental or practical fails to deliver. Preferring optimization techniques makes it cheaper and more economical. Optimizing process parameters can help a part to reach the required level of quality.

Optimization technique is popular in many areas around the world and it is also increasingly used in education as well as research purposes. Moreover, it is adopted by occasional users because even for the complex geometry of materials to weld it delivers tangible results in a short period. Techniques for optimizing process parameters in explosive welding include: Finite Element method (FEM) and Machine Learning (ML). The FEM method requires specific

Content from this work may be used under the terms of the Creative Commons Attribution 3.0 license. Any further distribution of this work must maintain attribution to the author(s) and the title of the work, journal citation and DOI. Published under license by Materials Research Forum LLC.

Materials Joining and Manufacturing Processes: MJMP 2025   Materials Research Forum LLC
Materials Research Proceedings 55 (2025) 51-56   https://doi.org/10.21741/9781644903612-9

information whereas ML requires a large amount of data [8]. Therefore, the current study reviews the influence of optimization parameters on the explosive welding method.

Algorithms in artificial intelligence (AI) have demonstrated significant promise in welding process parameter optimization and the data-driven have made it possible for the study. The use of AI algorithms to optimize welding process parameters is the main focus of the present paper. Using a variety of input variables, including kind of material, joint design, and weld quality, a machine learning model will be created that precisely forecasts the ideal welding parameters.

Actually, in AI, a dataset of previous welding tests will be used, where the parameters and associated weld quality will be noted. These findings are important because they have the potential to transform the welding industry by increasing the efficacy and efficiency of the welding process. The AI model can swiftly evaluate and pinpoint the connections between welding parameters and input components by utilizing AI algorithms, producing recommendations for the best parameters. So, it may lead to fewer welding flaws, better weld quality, and more output [9–11].

It is reported that to determine the influencing parameters, a two-level three factorial design is used to optimize the process parameters in the explosive cladding of titanium stainless steel 304L plates. In the study, analysis of variance was used to determine the linear, regression, and interaction values including the amplitude as well as the wavelength of the way interfaces were estimated using mathematical models. It is reported that concurrent with design predictions, the microstructure of the Ti-SS304L explosive-clad contact exhibited distinctive undulations [12].

In other findings, the Taguchi technique was used by the researchers to determine the important EXW process parameters such as standoff distance, loading ratio, and preset angle—in order to improve the aluminium-copper explosive clads' tensile strength. Three-level Taguchi L9 orthogonal arrays were used as the basis for the investigations. To forecast the tensile strength of Al-Cu explosive clads at a 95% confidence level, an empirical relationship based on multiple regression analysis was also created. The findings indicated that a higher tensile strength of Al-Cu explosive clad is strongly influenced by standoff distance, loading ratio, and preset angle [13].

Authors determined the tensile strength of explosively cladded aluminum-Stainless steel utilizing S/N ratio analysis based on the three factor-three level L9 Taguchi design. It was observed that at a 95% confidence level, the suggested mathematical model using the ANOVA method accurately forecasted the tensile strength of Al-SS304 clads [14–15].

A mathematical relationship between the welding input parameters and the weld joint's output variables was developed using an application of design of experiment (DoE), evolutionary algorithms, and computational networks to determine the desired weld quality [16]. A detailed review of the application of DoE methods in the area of welding was carried out by the authors. It was concluded that the high level of interest in the adaptation of surface methodology (RSM) and artificial neural networks (ANNs) to predict response(s) and optimize the welding process is essential and combining Genetic algorithm (GA) and RSM optimization techniques would reveal good results for finding out the optimal welding conditions [16].

A review of the influence of experimental parameters on the mechanical as well as microstructural characteristics of the resultant welded composite materials and the challenges encountered during the joining process are presented by the authors [17]. A special focus on underwater explosive welding was made. It was stated that by operating near the lower boundaries of the weldability window (where energy levels are reduced), welding defects such as porosity, cracks, and undesirable melted regions can be minimized. Also, by placing interlayers kinetic energy loss and the formation of IMC layers can be managed. The authors concluded that the EXW requires precise control on the thickness of the flyer plate, stand-off distance (SoD), loading ratio, the usage of intermediate layers, and detonation velocity for optimum results and desired mechanical and micro structural properties [17].

Materials Joining and Manufacturing Processes: MJMP 2025   Materials Research Forum LLC
Materials Research Proceedings 55 (2025) 51-56   https://doi.org/10.21741/9781644903612-9

The above-cited research review suggested that the adoption of AI algorithms to optimize the welding process parameters has been increasingly popular in recent years. The potential of AI In welding processes predicts weld quality, boosts production, and lowers costs. The generalizability of AI-based models across numerous welding circumstances and materials requires investigation.

Methodology

In this step, the information related to welding process parameters and the welding quality features will be gathered. Welding experiments, old documents, or internet databases can all provide this information. Variables including welding current, voltage, travel speed, shielding gas flow rate, weld bead geometry, and flaws may be included in the data that is gathered. To guarantee compliance with the AI algorithms, the data will be pre-processed to handle missing values, and outliers to normalize the characteristics. The flow chart indicated in Figure 1, provides information on how to apply the algorithm to predict the optimum parameters to obtain better bondability in EXW process [9].

| Feature Selections | ➡ | AI Algorithm | ➡ | Model Training | ➡ | Evaluation |

Fig 1: Workflow for Machine Learning Model Development [9]

Future selection: For the process of modelling, a subset of pertinent features will be chosen based on feature analysis and domain expertise. To create new features for the correlations between the welding process parameters and the intended welding results, engineering techniques may be used. The AI algorithm's performance may be enhanced by adopting engineered qualities.

AI Algorithm Selection and Architecture: Numerous artificial intelligence systems, including support vector machines, neural networks, genetic algorithms (RA), and random forests (RF), can be taken into consideration for the optimization of welding process parameters. The particular goals of the study and the characteristics of the dataset will determine which AI algorithm is supposed to be used. The number of layers, neurons, activation functions, and optimization techniques of the selected AI algorithm's architecture will be specified.

Model Training and Evaluation: Sets of training and testing will be created from the collected and pre-processed data. By taking into account of the optimization goal, the AI system will be trained on the training set using a suitable learning algorithm. In order to reduce the prediction errors, the model will be iteratively improved by modifying the parameters of the algorithm. The prediction performance of the trained model will be assessed using the testing set. Several performance indicators can be used to assess, how well the AI model predicts the welding results. Depending on the particulars of the issue being addressed, these measurements could include mean squared error, accuracy, precision, recall, or F1 score [12]. Based on the specified welding process parameters, these metrics will offer quantitative assessments of the model's capacity to forecast the intended welding quality features [9].

How well does the AI system predict: It is necessary to assess how well the AI system predicts ideal welding process parameters and how this affects the aspects of welding quality. Model performance can be evaluated using a variety of metrics, including accuracy, precision, recall, and F1 score. A thorough analysis and discussion of the AI algorithm's recommended ideal welding process parameters is necessary. The importance of the ideal welding parameters in reaching the intended welding results should be emphasized in the conversation.

Materials Joining and Manufacturing Processes: MJMP 2025   Materials Research Forum LLC
Materials Research Proceedings 55 (2025) 51-56   https://doi.org/10.21741/9781644903612-9

It is necessary to show a comparison between the AI algorithm and conventional techniques for choosing welding process parameters. The effectiveness, precision, and dependability of the AI algorithm can be compared to more conventional techniques like trial-and-error procedures or expert knowledge-based systems. The benefits of utilizing AI algorithms in terms of time, cost, and general performance should be highlighted in the conversation. It is important to discuss and acknowledge the study's shortcomings. These could include restrictions on the AI algorithm, the dataset, or the experimental design.

Future research directions could be suggested in order to get over these restrictions and enhance the use of AI algorithms in welding process parameter selection. This could entail adding more data sources, improving the AI algorithm, or investigating fresh methods for welding parameter optimization.

Figure 2a, b and c show the variations of normalised properties against wire speed in (m/min) in explosive welding. The normalised properties such as height, width, penetration, area and dilution with different colours against wire speed shown in Figure 2. The height, dilution and penetration decreases with increased wire speed. The width as well as area increase with increased wire speed. The normalised properties vary from 0.75 to 1 in Figure 2(a). The normalised properties in Figure 2(b) vary from 0.2 to 1 and all the properties start decreasing with increased wire speed till 4.375 m/min after that area increases significantly and height, width and dilution significantly will increase. Figure 2(c) shows that all the properties except dilution decrease with increased wire speed. As per this document coding is done in machine learning.

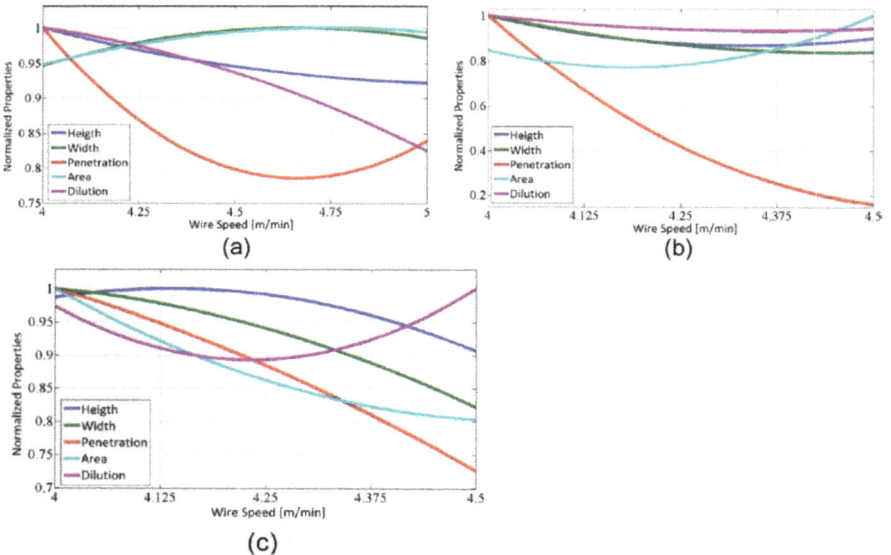

Fig 2: Analysis of AI Algorithm Impact on Welding Parameters [9]

Summary

Based on the literature survey, it was observed that a minimal effort had been made to understand the explosive welding behaviour of materials through Artificial intelligence machine learning

Materials Joining and Manufacturing Processes: MJMP 2025   Materials Research Forum LLC
Materials Research Proceedings 55 (2025) 51-56   https://doi.org/10.21741/9781644903612-9

(AIML) approaches. So in order to save time and money, and also to enhance the quality of the welding the use of AI algorithms and optimization techniques are highly essential. Though the review suggests the adaptation of surface methodology (RSM) and artificial neural networks (ANN)s to predict the response(s) and optimize the welding process parameters, it is recommended to use any two optimization techniques preferably Genetic algorithm (GA) and RSM and also the best-advanced simulation software to predict the best outcomes for the better optimal explosive welding conditions.

References

[1] Satyanarayan, S. Tanaka, A. Mori and K. Hokamoto, Welding of Sn and Cu Plates Using Controlled Underwater Shock Wave", J. Mater. Process. Technol. , 245 (2017) 300-308 https://doi.org/10.1016/j.jmatprotec.2017.02.030

[2] Satyanarayan, A. Mori and K. Hokamoto "Underwater shock wave weldability window for Sn-Cu plates" J. Mater. Process. Technol. 267 (2019) 152-158. https://doi.org/10.1016/j.jmatprotec.2018.11.044

[3] Satyanarayan, S. Tanaka and K. Hokamoto, Underwater Explosive Welding of Tin and Aluminium Plate Mater. Res. Proc, 13 (2019) 149-153. https://doi.org/10.21741/9781644900338-25

[4] Satyanarayan, K. Hokamoto, S. Tanaka, A Mori, D. Inao, The effect of interfacial morphology and weldability window on tin and aluminium plates welded plates using regulated water Shockwaves", Welding in the World, 68 (2024) 2941-2951. https://doi.org/10.1007/s40194-024-01834-1

[5] A. Gheysarian, M. Honarpisheh, Process Parameters Optimization of the Explosive-Welded Al/Cu Bimetal in the Incremental Sheet Metal Forming Process, Iran J. Sci. Technol.-Trans. Mech., 43 (2018) 945-956. https://doi.org/10.1007/s40997-018-0205-6

[6] A. Mustafa, B. Gulunec, F Findik, Investigation of explosive welding parameters and their effects on microhardness and shear strength, Mater. Des. 24 (2003), 659-664. https://doi.org/10.1016/S0261-3069(03)00066-9

[7] Z. Lu, L. Guo, L. Haung, J. Chen, Anisotropy in microstructure and shear properties of TA2/Q235 explosive welding interfaces, JMR&T, 25(2023). https://doi.org/10.1016/j.jmrt.2023.07.080

[8] I. Baturynska, O. Semeniutas, K Martinsen, Optimization of Process Parameters for Powder Bed Fusion Additive Manufacturing by Combination of Machine Learning and Finite Element Method: A Conceptual Framework, Procedia CIRP 67(2018), 227-232. https://doi.org/10.1016/j.procir.2017.12.204

[9] V. Dogra, Application of Welding Process Parameters Using AI Algorithm, Turk. J. Comput. Math. Educ, 9 (2018) 632-642. https://doi.org/10.17762/turcomat.v9i2.13866

[10] S. Yaknesh, N. Rajamurugu, P.K Babu, et al. A technical perspective on integrating artificial intelligence to solid-state welding. Int J Adv Manuf Technol 132, (2024) 4223-4248. https://doi.org/10.1007/s00170-024-13524-9

[11] Saravanan S, Kumararaja K. Raghukandan K. Application of Deep Learning Techniques to Predict the Mechanical Strength of Al-Steel Explosive Clads. Metals, 13 (2023), 373. https://doi.org/10.3390/met13020373

[12] P. Tamilchelvan, K. Raghukandan K, S. Saravanan, Optimization of process parameters in the explosive cladding of titanium/stainless steel 304L plates. Int J Mater Res. 104 (2013) 1205-1211 https://doi.org/10.3139/146.110977

[13] Elango E, Saravanan S and Raghukandan K, Optimization of process parameters in dissimilar explosive cladding through taguchi method, J. Manuf. Eng., 10 (2015), 4, 194-199.

[14] G. Costanza, E T Maria, C. Diego C, Explosion Welding: Process Evolution and Parameters Optimization, Mater. Sci. For., 941 (2018) 1558-64. https://doi.org/10.4028/www.scientific.net/MSF.941.1558

[15] D C Wilson, S. Saravanan, K. Raghukandan Optimization of process parameters in multilayer explosive cladding using taguchi method, JME , 11 (2016) 2, 107-112.

[16] K.Y Benyounis, A G Olabi, Optimization of different welding processes using statistical and numerical approaches - A reference guide, Adv. Eng, 39 (2008) 6, 483-496. https://doi.org/10.1016/j.advengsoft.2007.03.012

[17] B B Sherpa, R Rani, Advancements in explosive welding process for bimetallic material joining, A review, JALMES 6 (2024) 100078. https://doi.org/10.1016/j.jalmes.2024.100078

Materials Joining and Manufacturing Processes: MJMP 2025   Materials Research Forum LLC
Materials Research Proceedings 55 (2025) 57-63   https://doi.org/10.21741/9781644903612-10

An investigation of thermogravimetric analysis and thermal conductivity of glass fibre epoxy resin composites modified with silicon carbide, manganese, and copper nanoparticles (NPs)

Gurushanth B. Vaggar[1,a*], S.C. Kamate[2,b], Kiran C.H.[1,c], Deepak Kothari[1,d], S.L. Nadaf[3,e], Vishalagoud Patil[3,f]

[1]Department of Mechanical Engineering, AIET (Affiliated under VTU, Belagavi), Moodbidri 574225, India

[2]Hirasugar Institute of Technology, Nidasoshi, India

[3]Mechanical Engineering Department GEC, Talakal, India

[a]gvaggar7@aiet.org.in, [b]kamateksk@rediffmail.com, [c]ckmkiran@aiet.org.in, [d]deepakkothari@aiet.org.in, [e]slnadaf29@gmail.com, [f]vishalspatil2008@gmail.com

Keywords: Silicon Carbide (SiC) Nanoparticles, Glass Fibre (GF), Epoxy Resin (ER), Manganese (Mn) Nanoparticles, Copper (Cu) Nanoparticles, Thermal Conductivity, Thermogravimetric Analysis (TGA)

Abstract. In the current work, the effect of copper (Cu), manganese (Mn), and silicon carbide (SiC) nanoparticles on the thermal stability and conductivity of glass fiber-epoxy resin composites was discussed. Thermogravimetric Analysis (TGA) was done as per ASTM E 1131 to study the thermal degradation of composites from 30°C to 800°C. Thermal conductivity increases with the percentage of fillers of SiC, Mn, and Cu nanoparticles in the GFER polymer composites. The combination of epoxy, Glass Fibre and SiC nanoparticle-loaded composites was found highest thermal conductivity. TGA discloses that for SiC, Mn and Cu NPs glass fibre epoxy resin composites thermal stability increased with the percentage weight fraction of nanoparticles. The addition of nanoparticles to glass fibre epoxy resin polymer composites causes an increase in thermal properties due to the nanoparticles acting as catalysts for decomposition and less mass loss observed in TGA. The Thermal Conductivity of 20% SiC nanoparticles were found to be 0.37 W/mK. In Overall comparison, SiC nanoparticle-loaded composites exhibit significantly improved thermal stability and are more suited where the heat transfer plays the predominant role in high thermal conductivity and more thermal stability application fields.

Introduction
The annual rise in demand for GFER composites within manufacturing sectors drives the pursuit of more efficient composite materials that exhibit superior properties compared to their individual components. The hybrid reinforced composites are the materials to prepare the required property-oriented one. In Composites, fibre increases the strength, stiffness and thermal stability and resin helps in aligning fibres by distributing stress uniformly. Excellent electrical resistance, low maintenance needs, non-corrosiveness, and a high ratio of strength to weight are just a few of its many advantages. Composites are manufactured through various techniques that depend on the type of matrix, reinforcement, and intended application. Among the diverse methods employed are pultrusion, autoclaving, resin transfer moulding, manual layup, compression moulding, and filament winding [1].

The thermal property variations are nonlinear in hybrid composites, it is very difficult to identify and predict the temperature behaviour of these materials with unknown thermal conductivity. However, thermal conductivity measurement of hybrid composite materials is not an easy task due to temperature variations complexity [2]. The addition of Silicon carbide filler particles to neat

Content from this work may be used under the terms of the Creative Commons Attribution 3.0 license. Any further distribution of this work must maintain attribution to the author(s) and the title of the work, journal citation and DOI. Published under license by Materials Research Forum LLC.

Materials Joining and Manufacturing Processes: MJMP 2025   Materials Research Forum LLC
Materials Research Proceedings 55 (2025) 57-63   https://doi.org/10.21741/9781644903612-10

epoxy polymer composites increases thermal conductivity as the volume fraction percentage increases [3]. Hybrid nano-filler-modified polymer matrices have created wide opportunities in research to achieve excellent thermal and mechanical properties for advances in many fields of applications. The commonly used nanoparticle-fillers are carbon nanoparticle-filler, carbon nanotubes, nano clay and silicon carbide fibres [4]. Since polymers have low thermal conductivity, an effort has been made to increase the thermal conductivity and electrical resistivity of polymers by adding boron nitride, silicon carbides, glass Fibre, mica, aluminium nitride, alumina and zinc oxide fillers [5]. Improved Fire resistance and more thermal stability is found in silica hybrid composites, due to the high strength of interfacial bonding of silica particles and epoxy resin [6]. The addition of graphene nano-plates (1% weight) increased thermal conductivity and the enhancement was significant due to the homogenous dispersion of graphene nano-plates in the epoxy polymer matrix and internal bonding. Thermal conductivity improvement is 33.3% for 1% graphene nano-plates compared to plain epoxy resin [7].

Both organic and inorganic materials are present in the epoxy resin-glass fibre composite, and the glass fibre coating controls the interface between the matrix and the fibre [8]. Thermally and electrically insulated polymers are frequently used in electrical and electronic equipment. Heat dissipation becomes a significant issue with electrically insulating polymer materials, and poor thermal conductors [9]. Since nearly all engineering specialities employ polymer matrix composites, it is imperative to enhance these materials thermal characteristics [10]. Poor heat conductivity, poor chemical and environmental stability, and poor heat resistance are some of the drawbacks of planar polymer materials. To solve these problems, new filler particles need to be incorporated into the composite designs. [11]. The primary purpose of additives is to change and fine-tune the composition's material properties [12]. Low heat conductivity, poor chemical and environmental stability, and poor heat resistance are some of the drawbacks of planar polymer materials. A novel solution for excellent heat resistance and stability in varying temperatures is polymer composites (PCs) [13]. The current study focused on improving the thermal properties of glass fibre epoxy resin composites by adding nanoparticles.

Production of Polymer Composites
Polymer matrix-modified composites are developed using two distinct composition calculation techniques: Method 1: Establish a direct ratio of glass fibre (GF) to epoxy resin (ER) within the range of 40:60 to 50:50, utilizing weight fractions of nanoparticle fillers that vary from 5% to 30%. Method 2: The preparation of composites is influenced by the thickness of the specimens, which determines the weight fraction of the fibre, epoxy resin, hardener, and nanoparticle filler utilized. The following polymer matrix-modified composite compositions were attained by using the second method and the designation of specimen samples shown in Table 1.

Experimental Setups: Thermal Conductivity Setup
Materials whose thermal conductivity falls between 1 W/mK and 10 W/mK are tested using the FOX 50 heat flow meter which works on the steady state heat flow principle. The device exhibits rapid results, a lightweight, and user-friendly instrument for measuring thermal conductivity (Figure 1). It works best with optical encoders that can operate in a varied temperature range when measuring the specimen's thickness digitally. WinTherm-50 software regulates the device, which has solid-state heating and cooling and thin-film heat flux transducers to measure heat flow. In each part of the heat conductivity formula ($K = (\Delta X_{specimen} / R_{specimen}) = \Delta X_{specimen} / (R_{total} - 2R_{contact\ plates})$), the thermal conductivity of the modified composite material can be measured using the specimen's thickness, the temperature differential across it, and the steady state heat flux, The Thermal Conductivity test was conducted at 60°C, ΔT:20°C, Top Plate-70°C, Bottom Plate-50°C, Average Temperature-60°C.

Materials Joining and Manufacturing Processes: MJMP 2025   Materials Research Forum LLC
Materials Research Proceedings 55 (2025) 57-63   https://doi.org/10.21741/9781644903612-10

Table 1. Composition designation of composites made of SiC, Cu, and Mn, nanoparticles.

Designation	Designation	Designation
GF	GF	GF
GFSiC-NPs5	GFMn-NPs5	GFCu-NPs5
GFSiC-NPs10	GFMn-NPs10	GFCu-NPs10
GFSiC-NPs15	GFMn-NPs15	GFCu-NPs15
GFSiC-NPs20	GFMn-NPs20	GFCu-NPs20

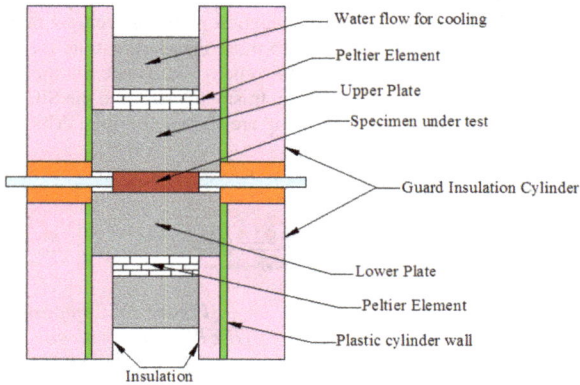

Figure 1. *FOX 50 Heat flow meter cross-section view.*

Thermogravimetric analysis (TGA) setup.

Samples in TGA are exposed to a varied range of temperatures, and their behaviour is tracked over time at different temperature points. The TGA tests were performed at Bengaluru, at CIPET (School for Advanced Research in Polymers). Samples that are square and rectangular in shape and weigh between 5 and 10 g are used in TGA.

All processed data related to these tests were recorded during the test and tabulated appropriately. TGA test data is provided by the computer-controlled STA 7300 and TGA 8000 TGA equipment. Temperature variations over time, alterations in sample weight loss as a function of temperature, and modifications in the flow parameters of oxygen and nitrogen are all directly influenced by time. The rate of temperature of STA 7300 TGA test heating chamber is 20 °C per minute in three steps when nitrogen is supplied at 50 ml/min, as Table 2 (a, b) illustrates.

Table 2a. *TGA test nitrogen flow.*

Thermal degradation phases	Nitrogen flow	Rate of rise in temperature
I - Phase	50 millilitre/minute	20°C per minute (up to 300°C)
II - Phase	50 millilitre/minute	20°C per minute (up to 600°C)
III- Phase	50 millilitre/minute	20°C per minute (up to 900°C)

Table 2b. *TGA test oxygen flow.*

Thermal degradation phases	Oxygen flow.	Rate of rise in temperature
I - Phase	50 millilitre/minute	20°C per minute (up to 300°C)
II - Phase	50 millilitre/minute	20°C per minute (up to 600°C)
III- Phase	50 millilitre/minute	20°C per minute (up to 900°C)

Materials Joining and Manufacturing Processes: MJMP 2025

Materials Research Proceedings 55 (2025) 57-63

Materials Research Forum LLC

https://doi.org/10.21741/9781644903612-10

Results and Discussion

Phonons are known to be a significant factor in finding the thermal conductivity, of GFER polymer composite materials, which has been simultaneously altered by phonon diffusion through boundaries at low temperatures and by both harmonic and anharmonic phonon interactions at high temperatures. Pure fibreglass epoxy resins have a low degree of structural order and phonons interacting with other phonons from randomly entangled molecular chains prevent it from achieving the desired thermal conductivity. A complete heat transfer pathway would be formed if the number of SiC NPs is increased further and it crosses the percolation threshold, creating a bridge between the local chain and network. Modified GFER polymer composite's thermal conductivity is enhanced by the creation of conduction path connections that aid in producing continuous phases by GFER SiC NPs. In order to more clearly illustrate the superiority of SiC-NPs, Figure 2(a) presents a model of the fibreglass epoxy composite thermal conductivity. High thermal interface resistance is produced by phonon mismatch between the SiC-NPs and the GFER matrix. The heat conduction pathways after heating are shown in Figure 2(b).

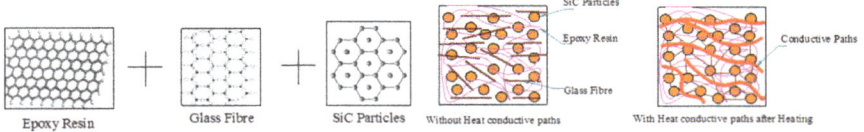

Figure 2a. GFER Polymer Composites with SiC Particle Filler Formulation.

Figure 2b. Composites made of GFERPolymer with and without thermal conduction paths filled with SiC particle filler.

The Cu nanoparticle composites have shown lower thermal conductivity than SiC nanoparticle composites because the high thermal conductivity of Cu allows molecules to rapidly separate, leading to the decomposition of polymer chain links. Figure 2(c, d) shows the mechanism of the heat conduction pathways following the heating of the modified polymer composites.

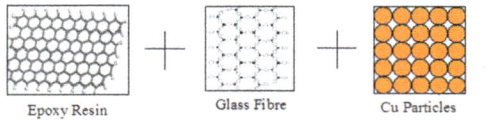

Figure 2c. Copper Particle Filler GFER Polymer Composites Formulation.

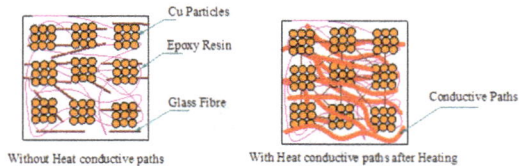

Figure 2d. GFER Polymer Composites with and without Thermal Conduction Paths for Copper Particle Filler.

The nanoparticles improve the phonon scattering, which causes better heat pathways, and increases the thermal conductivity. The weight percentage of fillers added to SiC, Mn, and Cu-

Materials Joining and Manufacturing Processes: MJMP 2025   Materials Research Forum LLC
Materials Research Proceedings 55 (2025) 57-63   https://doi.org/10.21741/9781644903612-10

modified composites increases their thermal conductivity (Figure 3). When compared to copper and manganese compounds, SiC compounds exhibit a higher thermal conductivity. This research work will contribute to the development of modified polymer composites with high thermal conductivity by using different fillers based on specific field needs and applications. Material selection depends on factors like weight, strength, stiffness, and heat resistance. Because of this, strong, low-density, and heat-resistant advanced materials will be used in various applications in the future.

The findings drawn from Figures 4(a), 4(b), and 4(c) indicate that in comparison to pure GFER composites, silicon nanoparticulate thermal stability better at filler contents of 15% and 20%, while copper nanoparticulate composites exhibit inferior thermal stability. Manganese and copper NPs composites performed worse than SiC NPs composites. The crosslink density of epoxy and its compounds is correlated with the temperature at which a material transition occurs (Tg). Due to the uneven degree of cross-linking in the epoxy matrix, glass transition temperatures happen over a broad temperature range. Consequently, Tg's typical temperatures can be found at the starting temperature, intermediate temperature, and final temperature. The addition of SiC NPs to the fibreglass epoxy resin matrix raises the Tg, suggesting that interactions occur between the matrix molecules and the strong interface during the curing process. As a result, there are more barriers to preventing macromolecular chain motion, raising the Tg and encouraging thermal stability.

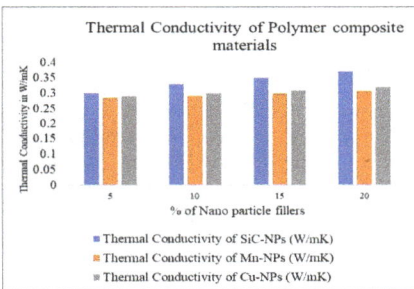

Figure 3. Thermal conductivity comparison.

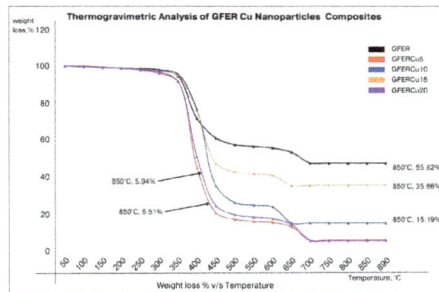

Figure 4a. TGA of Cu NPs Composites.

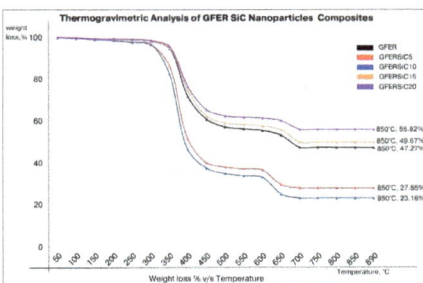

Figure 4b. TGA of SiC NPs Composites.

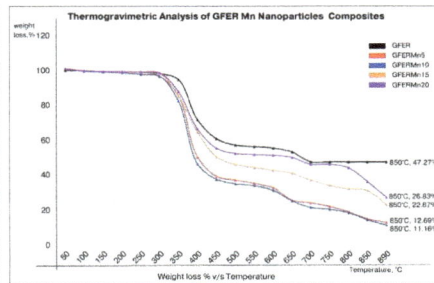

Figure 4c. TGA of Mn NPs Composites.

Conclusions

- For 15%–20% silicon carbide nanoparticle-modified composites, there is a 9 to 25% increase in thermal conductivity. Not significantly improved over pure GFER composites at low percentages (5 percent and 10 percent) of SiC, Mn, and Cu nanoparticle composites.

- TGA indicates that silicon carbide nanoparticle polymer composites have higher thermal stability than copper and manganese nanoparticle polymer composites. When compared to pure GFER composites, Mn and Cu nanoparticle composites do not significantly improve thermal stability, although composites containing 15 and 20 percent SiC nanoparticles exhibit higher thermal stability.

- Mn and Cu nanoparticle composites do not significantly improve thermal stability when compared to pure GFER composites, although higher thermal stability is noted in composites containing 15% and 20% SiC nanoparticles (Figure 4 a-c). In contrast to manganese and copper nanoparticle filler compounds, silicon nanoparticle filler compounds exhibit superior thermal stability due to the lower temperature at which copper nanoparticle filler compounds decompose.

- The weight percentage results of thermogravimetric analyses using the STA 7300 and TGA 8000 show that loss in Mn and Cu nanoparticle-modified composites is higher than loss in SiC nanoparticle-modified composites.

References

[1] Y. L. Brian, O. Jason, Properties of composite materials for thermal analysis involving fires, Comp. part A 37 (2006) 1068-1081. https://doi.org/10.1016/j.compositesa.2005.01.029

[2] K. Dilek, H. T. Ismail, M. T. Coban, Thermal conductivity of particle filled polyethylene composite materials, Comp. Sci. and Tech. (2003) 113–117, https://doi.org/10.1016/S0266-3538(02)00194-X

[3] S. Krishnamachar, M. B. Siddalingappa, Thermal conductivity enhancement of epoxy by hybrid particulate fillers of graphite and silicon carbide, J of Min. and Mat. Charact. and Engg., (2015), 3, 76-84. http://dx.doi.org/10.4236/jmmce.2015.32010

[4] J. Naveen, Md. Jawaid, Z. Edi Syams, Md. H. S. Thariq, Y. Ridwan, Enhanced thermal and dynamic mechanical properties of synthetic/natural hybrid composites with graphene nanoplateletes, Polymers (2019), 11, 1085. https://doi.org/10.3390/polym11071085

[5] C. Watthanaphon, F. Daisuke, T. Shuichi, U. Hideyuki, I. Yoshiyuki, Thermal and mechanical properties of polypropylene/boron nitride composites, Energy Procedia 34 (2013) 808 – 817. https://doi.org/10.1016/j.egypro.2013.06.817

[6] D. Kim, I. Chung, G. Kim, Study on Mechanical and thermal properties of fiber-reinforced epoxy/hybrid-silica composite, Fib. and Poly. (2013), 2141-2147. https://doi.org/10.1007/s12221-013-2141-9

[7] M. T. Le, S. C. Huang, Thermal and mechanical behaviour of hybrid polymer nanocomposite reinforced with graphene nano platelets, Materials (2015), 8, 5526-5536, https://doi.org/10.3390/ma8085262

[8] S. R. Akm, R. Vijaya, S. Jeelani, Thermal and mechanical properties of woven glass fiber reinforced epoxy composites with carbon nanotubes grown in-situ, IJES, ISSN (e): 2319 – 1813 ISSN (p): 2319 – 1805 (2015).

[9] A. K. Aseel, Enhanced thermal and electrical properties of epoxy/carbon fiber–silicon carbide composites, Adv. Comp. Letrs. (2020), https://doi.org/10.1177/2633366X19894598

[10]B. Bhasker, Dr. M. Devaiah, P. Laxmi Reddy, M. R. Gandhi, Thermal characterization of fibre reinforced polymer composites and hybrid composites, IJME&T, Volume 10, Issue 03, (2019), pp. 1055–1066, Article ID: IJMET_10_03_106.

[11]G. Mittal, K. Y. Rhee, M. S. Vesna, D. Hui, Reinforcements in multi-scale polymer composites: Processing, properties, and applications, Comp. Part B 138 (2018) 122-139, https://doi.org/10.1016/j.compositesb.2017.11.028

[12]R. Ambigai, S. Prabhu, Analysis on mechanical and thermal properties of glass carbon/epoxy based hybrid composites, Matls. Sci. and Engg. 402 (2018) 012136. https://doi.org/10.1088/1757-899X/402/1/012136

[13]D. H. Ebadi, M. N. S. Branch, Thermal conductivity of nanoparticles filled polymers, Islamic Azad UISNT. 519-540 (2012). https://doi.org/10.5772/33842

Materials Joining and Manufacturing Processes: MJMP 2025   Materials Research Forum LLC
Materials Research Proceedings 55 (2025) 64-71   https://doi.org/10.21741/9781644903612-11

A review on effect of filler materials on thermal properties of hybrid polymer matrix composites

Gurushanth B. Vaggar[1,a*], S.L. Nadaf[2,b], Narayan V.[1,c],
Dharshith[1,d], Praveen V.[1,e]

[1]Department. of Mechanical Engineering, Alva's Institute of Engineering and Technology, Moodbidri. (Affiliated to VTU Belagavi, India)

[2]Department. of Mechanical Engineering GEC, Talakal, India

[a]gvaggar7@aiet.org.in, [b]slnadaf29@gmail.com, [c]narayanv628@gmail.com,
[d]laludharshith@gmail.com, [e]praveenvc321@gmail.com

Keywords: Glass Fibre, Epoxy Resin Composites, Filler Particles, Thermal Conductivity, Thermogravimetric Analysis, Coefficient of Thermal Expansion

Abstract. Hybrid polymer composites are increasingly in demand due to their lightweight and high-strength properties. However, traditional polymers exhibit poor thermal properties and limited strength under high-temperature conditions. Glass fibre-reinforced epoxy composites, offer excellent strength and stiffness, with poor thermal stability, leading to degradation under heat. Enhancing the thermal properties of these materials without compromising mechanical integrity is critical. Incorporating high thermal conductivity filler particles, such as CNTs, graphene, silicon carbide, copper powder, or copper mesh, into carbon fibre-reinforced polymer (CFRP) and glass fibre-reinforced polymer (GFRP) creates hybrid composites with improved thermal stability and resistivity. These hybrid composites can withstand high-temperature environments while maintaining structural strength and stiffness, broadening their applications across aerospace, automotive, pressure vessels, and wind energy industries. These sectors collectively account for 52.17% of the total glass fibre market. Globally, the demand for glass fibre is 47.80 million tons per year, growing annually at a rate of 10.7%. Despite the challenge of nonlinear thermal property variations and complex temperature behaviour, the strategic addition of fillers significantly improves thermal performance. This makes polymer composites a vital material in advanced structural applications where both thermal resistance and mechanical reliability are paramount. The overview of the current work is to study the effect of filler particles on the thermal properties of polymer composites.

Introduction

Composite materials are advanced structures formed by combining two or more distinct materials, each retaining its unique properties while contributing to the overall performance of the composite. These materials are characterized by a discrete interface that separates them, resulting in a heterogeneous structure. Of all types of composites, synthetic composites stand out because they possess great strength and stiffness-to-weight ratios, thereby allowing them to be used for high-performance structural applications. Based on this characteristic, the term composite can be further narrowed down to include only composites containing high-strength Fibre reinforcements supported by a strong matrix material, either organic or inorganic. The performance of a composite is often significantly influenced by the fibre reinforcement properties; increased volume fractions of fibres often lead to improved mechanical properties until the point where the matrix can no longer provide sufficient support to the fibres. Classifying fibre-reinforced composites primarily involves consideration of the shape of the fibrous reinforcement: these may include continuous, long discontinuous, or short discontinuous fibres. Continuous and long fibre-reinforced

Content from this work may be used under the terms of the Creative Commons Attribution 3.0 license. Any further distribution of this work must maintain attribution to the author(s) and the title of the work, journal citation and DOI. Published under license by Materials Research Forum LLC.

Materials Joining and Manufacturing Processes: MJMP 2025   Materials Research Forum LLC
Materials Research Proceedings 55 (2025) 64-71   https://doi.org/10.21741/9781644903612-11

composites translate fibre properties maximally into composite performance, while short fibres lead to poor mechanical properties. The main drive behind the development of high-performance composites, especially those with resin matrices, is from industries that have an interest in using lighter-weight materials rather than traditional materials like aluminium and steel [1].

Optical fibre-reinforced composites are known to be cost-effective and relatively strong compared to other polymer matrix composites. In GFRP, the presence of silica (SiO_2) can improve performance while different types of glass fibres used, such as S-glass and E-glass, provide specific properties for certain applications. The high cost of polymers restricts the commercial application of polymers, but fillers can improve composite properties while reducing overall cost. GFRP composites find extensive applications in various industries, especially in marine and aerospace sectors, because of their excellent environmental resistance, damage tolerance, and specific strength. The dynamic stability of these composites under varying temperature conditions needs to be understood to optimize their performance in practical applications [2].

Among the most striking advancements are nanocomposites, where the infusion of nanoparticles—like carbon nanotubes and graphene—into polymer matrices creates materials with astonishing mechanical strength, thermal stability, and electrical conductivity. These nanocomposites are revolutionizing high-performance sectors such as aerospace and electronics, pushing the boundaries of what materials can achieve [3]. In an era increasingly concerned with environmental impact, engineers are championing the development of bio-based polymers. By harnessing renewable resources, they are transforming natural fibres—such as hemp, flax, and bamboo—into robust composite materials. This reduces dependence on fossil fuels and enhances biodegradability, paving the way for a more sustainable future [4].

Additionally, the integration of smart composites is revolutionizing material design. Engineers are embedding sensors and actuators into polymer matrices, creating materials that respond dynamically to environmental changes [5]. Beyond technological prowess, engineers are advancing environmental stewardship. By optimizing alloy compositions and manufacturing processes based on solubility limits, they forge pathways toward sustainable practices, enhancing material recyclability and minimizing environmental impact [6].

The Impact of Filler Particles on Thermal Properties of Polymer Composite Materials

A study shows characterized polylactic acid (PLA)/titanium dioxide (TiO_2) nanocomposites to evaluate their antibacterial potential. The variables considered were TiO_2 particle size (21 nm and <100 nm) and content (0%, 1%, 5%, 10%, 20%, wt %). Structural analysis showed no significant changes in the PLA matrix due to the nanoparticles. Thermal transitions, including glass transition, cold crystallisation, and melting, occurred at similar temperatures, while thermal degradation (Tg) temperatures slightly increased with TiO_2 content. The TiO_2 nanoparticles reduced bacterial growth and biofilm formation, decreasing extracellular polymeric substance (EPS) [7]. The research considered epoxy and epoxy nanocomposites with varying weight percentages of hexagonal boron nitride nanoparticles (BNNPs) (0.5, 0.7, 1.2, and 5%). The thermal stability of the nanocomposites improved as determined through TGA, DTG, and DSC analysis. Thermal stability was higher with increased weight fractions of boron nitride nanoparticles. It has been observed that there is a linear relationship between activation energy and activation entropy, which means that the addition of BNNPs enhances the thermal stability of the composites. This indicates that BNNP-reinforced composites have a higher resistance to thermal degradation [8].

The thermal stability of nanocomposites based on unsaturated polyester resin (UP) and Magnesium Oxide (MgO) nanoparticles (NPs) is enhanced according to TGA and DSC analysis. The TG analysis suggests that the weight fraction increase of MgO NPs enhances thermal stability. A linear relationship between activation energy and activation entropy indicates that the Coast-Redfern method is applicable to fit all experimental data. Furthermore, the enthalpy and entropy

Materials Joining and Manufacturing Processes: MJMP 2025   Materials Research Forum LLC
Materials Research Proceedings 55 (2025) 64-71   https://doi.org/10.21741/9781644903612-11

increase with the weight percentage of MgO, because it delays heat loss during oxidative decomposition, and this leads to the accumulation of heat within the specimens [9].

Epoxy resin reinforced Kevlar and glass fibre composites DSC test reveals that glass transition temperature (T_g) varies from 112.3°C to 120.2°C of different proportions (35:65, 40:60, 45:55, 50:50 ratio of Kevlar/Glass fibre) of Kevlar and glass fibre hybrid composites. Peak temperature sustains from 526.6°C to 555.4°C can be seen in figures 1a-d. In the DSC test, 5-6 mg specimens constantly heated at the rate of 10°C/min from 25°C to 580°C. Kevlar and glass fibre hybrid composite 10-15 mg specimens are used in TGA analysis almost all specimen's weight loss starts from around 300°C and the highest weight loss is between 550°C to 650°C. Maximum weight loss close to 800°C was 68.8%. For low % Kevlar found maximum weight loss, equal proportions of Kevlar and Glass fibre found less weight loss compared to other proportions [10]. Compared to plane composites with natural fillers, synthetic filler composites have enhanced the thermal properties such fillers play a vital role in electrical and electronics applications [11]. Adding more percentage of glass fibres in epoxy resin composites causes more thermal stability and an increase in glass transition temperature [12].

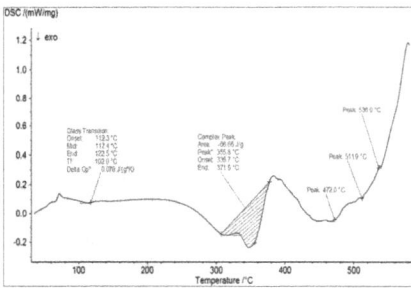

Figure 1a. DSC result of Kevlar/ Glass Fibre 35:65.[10]   Figure 1b. DSC result of Kevlar/ Glass Fibre 40:60. [10]

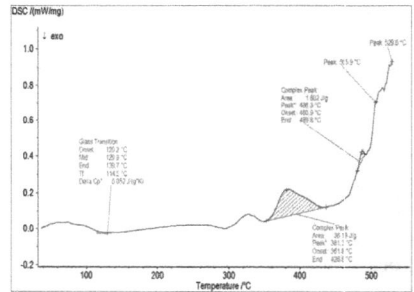

For the epoxy resin with rice husk powder composite, thermal conductivity decreases with the increase of rice husk filler percentage [13]. Figure 2a-c shows the Carbon/glass fibre hybrid composites exhibit more thermal stability and have high glass transition temperatures compared to individual fibres of carbon and glass and maximum weight loss was observed in glass fibre, less weight loss in hybrid fibre composites (Carbon/Glass) [14]. Thermal conductivity 10% Silicon carbide filler composites exhibit high thermal conductivity, almost double the neat epoxy Kenaf fibre (0% Silicon carbide) composites, due to high molecular movements of SiC. The weight loss was less and more thermal stability in 10% SiC filler composites and due to a delay in the degradation temperature of the composites while the interaction of SiC fillers with the composites [15].

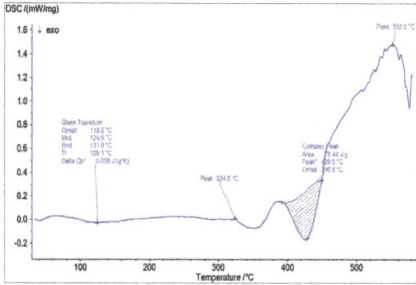

Figure 1c. DSC result of Kevlar/ Glass Fibre 45:55. [10]

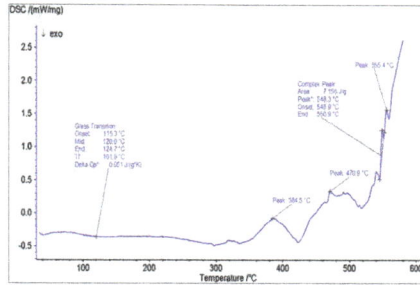

Figure 1d. DSC result of Kevlar/ Glass Fibre 50:50. [10]

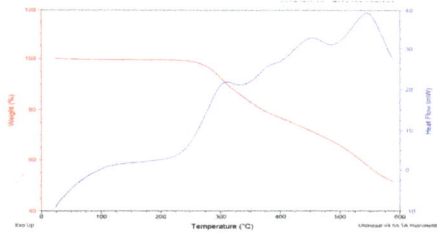

Figure 2a. DSC &TGA Curve for Carbon fibre composites. [14]

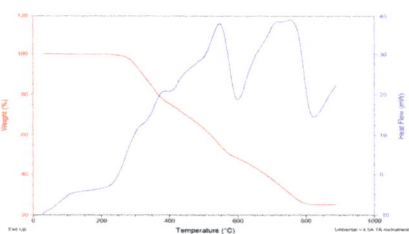

Figure 2b. DSC &TGA Curve for Glass fibre composites. [14]

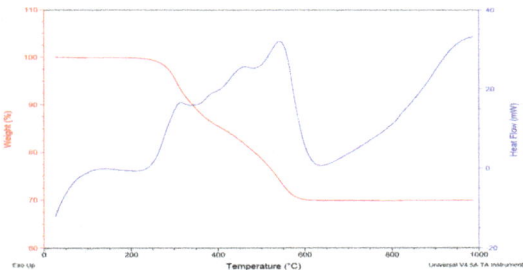

Figure 2c. DSC &TGA Curve for Hybrid (Carbon/Glass fibre) composites.[14]

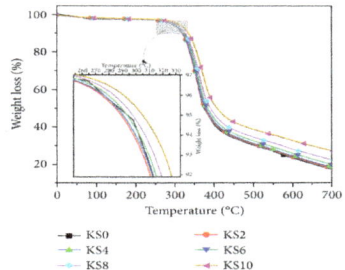

Figure 3. TG analysis of laminates (TGA).[15]

Polyoxymethylene (POM) glass fibre with Cloisite 25A nano clay (1%, 2%, 3%) composites decomposition temperature decreases with the increase of nano clay percentage, showing poor thermal stability. There is a slight increase in melting point POM with the addition of only Glass Fibre [16]. Glass fibre epoxy resin composites with graphite oxide fillers have exhibited good dielectric breakdown strength and dielectric constant values for electrical applications [17]. Glass fibre epoxy resin composite applications widely vary in all fields like, medical, educational, research, domestic household materials, toys, boats, electronics circuit boards, fibre roof sheets, chairs, aeroplanes, and automobiles [18]. With varying weight percentages of bulk moulding

Materials Joining and Manufacturing Processes: MJMP 2025   Materials Research Forum LLC
Materials Research Proceedings 55 (2025) 64-71   https://doi.org/10.21741/9781644903612-11

compounds in glass fibre epoxy resin composites from DSC analysis shows that specific heat increases with the increase of bulk moulding compounds percentage. In between the temperature range 320°C - 560°C, maximum weight loss was observed from TGA. Adding BMCs in glass fibre epoxy composites leads to a decrease in thermal conductivity [19]. In Polysulfone glass fibre composites the mechanical and thermo-mechanical properties improve with the increase of glass fibre to Polysulfone polymer ratio [20]. Adding Basalt fibres in Polyester resin glass fibre composites affects the dimensional stability of composites at higher temperatures and, the need to operate basalt glass fibre hybrid composites at lower temperatures [21]. Further Polylactic Acid Glass Fibre composites exhibit higher thermal stability than neat polylactic acid [22]. Polypropylene composites filled with rattan nanoparticles, onset Temperature and end temperature improve with the addition of rattan nanoparticles [23]. Table 1 provides a detailed information on effect of fillers on polymer matrix composites.

Table 1. Effect of fillers on polymer matrix composites.

Matrix/ Fibre	Fibre ratio	Fillers (wt%/Vol%)	Thermal Property Change	Refer ence
Polylactic acid (PLA)	----	Titanium dioxide (TiO$_2$) nanoparticles (0%, 1%, 5%, 10%, and 20%)	Thermal degradation temperatures slightly increased with TiO2 content	[7]
Epoxy	----	Hexagonal boron nitride nanoparticles (BNNPs) (0.5, 0.7, 1.2, and 5%)	The thermal stability of the nanocomposites improved as determined through TGA, DTG, and DSC analysis. Thermal stability was higher with increased weight fractions of boron nitride nanoparticles.	[8]
Unsaturate d polyester resin	----	Magnesium oxides (0, 1, 3, 5, 7, 10, 15%)	TGA, DTG and DSC analysis of nanocomposites shows enhancement in the thermal stability of nanocomposites, the TG will increase when the weight fraction of (MgO NPs) increases.	[9]
Epoxy / Kenaf Fibre	----	Silicon Carbide (2%, 4%, 6%, 8%, 10%)	Thermal conductivity of 10% Silicon carbide exhibits high thermal conductivity—weight loss less in 10% SiC filler composites and more thermal stability. Tg increases with an increase in SiC filler.	[15]
Polyoxyme thylene (POM) / Glass Fibre	30% GF & 70% POM	Cloisite 25A nanoclay (1%, 2%, 3%)	In TGA the decomposition temperature decreases with the increase of the percentage of nanoclay. Which shows poor thermal stability. There is a slight increase in melting point POM with the addition of only Glass Fibre.	[16]
Epoxy / Glass Fibre	----	Graphite Oxides (0.5%, 1%, 2.5%)	Dielectric characteristics are increased with the addition of graphite oxides.	[17]
Epoxy / Glass Fibre	----	Bulk Molding Compounds (BMCs) (0%, 5%, 10% 15%)	Maximum weight loss occurs between 320°C - 560°C temperature range. Cp value increases with BMCs addition intern thermal conductivity decreases.	[19]

Polysulfone / Glass Fibre	50:50, 60:40, and 70:30	----	With the increase of Glass fibre to Polysulfone ratio the Mechanical and Thermo-Mechanical properties improve.	[20]
Polyester resin / Basalt, Glass Fibre	70:30 (Polymer: Fibre)	----	From DMA test observed that the dimensional stability of composites was affected at higher temperatures, using basalt in a glass fibre matrix adversely affects the dimensional stability of composites. Therefore, Basalt glass fibre polymer composites are suited to lower-temperature applications.	[21]
Epoxy Resin / Glass Fibre	60:40	SiC Nanoparticles and Cu Nanoparticles (5%, 10%, 15% and 20%)	SiC-NPs composites have shown higher thermal conductivity values compared to Cu-NPs composites. More thermal stability was observed in 15% and 20% SiC-NPs composites	[24]

Conclusions

- The inclusion of nanoparticles in most cases leads to a noticeable improvement in the thermal properties of polymer composites.
- The enhancement in the thermal properties motivates researchers to further explore and develop hybrid polymer composites.
- The goal of the review work is to optimize the thermal properties of polymer materials based on the specific requirements of various applications.

References

[1] R. M. Christensen, Mechanics of Composite Materials, Wiley-Interscience, New York, (1979). 01-30.

[2] Md. A. Tabibzadeh, Tensile, compressive and shear properties of unidirectional glass/epoxy composites subjected to mechanical loading and low-temperature services, Ind. J. of Engg. & Mats. Sci., 20 (2013), 299-309.

[3] Y. Shi, J. Yang, F. Gao, Q. Zhang, Covalent Organic Frameworks: Recent Progress in Biomedical Applications. ACS Nano 17(3) (2023), 1879-1905. https://doi.org/10.1021/acsnano.2c11346

[4] M. H. Islam, S. Afroj, M. A. Uddin, D. V. Andreeva, K. S. Novoselov, N. Karim, Graphene and CNT-Based Smart Fiber-Reinforced Composites: A Review, Adv. Funct. Mater. 32, (2022), 2205723, 01-31. https://doi.org/10.1002/adfm.202205723

[5] E. El-Seidy, M. Sambucci, M. Chougan, M. J. Al-Kheetan, I. Biblioteca, M. Valente, S. H. Ghaffar, Mechanical and physical characteristics of alkali-activated mortars incorporated with recycled polyvinyl chloride and rubber aggregates. J. Build. Eng. 60, (2022),105043, 01-15. https://doi.org/10.1016/j.jobe.2022.105043

[6] A. G. Williams, E. Moore, A. Thomas, J. A. Johnson, Graphene-Based Materials in Dental Applications: Antibacterial, Biocompatible, and Bone Regenerative Properties. Int. J. Biomater. 8803283, (2023), 01-18. https://doi.org/10.1155/2023/8803283

Materials Joining and Manufacturing Processes: MJMP 2025
Materials Research Proceedings 55 (2025) 64-71

Materials Research Forum LLC
https://doi.org/10.21741/9781644903612-11

[7] A. S. G. Edwin, O. Dania, Á. L. Miguel, V. Itziar, G. B. Javier, Preparation and Characterization of Polymer Composite Materials Based on PLA/TiO2 for Antibacterial Packaging, Poly., 10, (2018), 1365, 01-14. https://doi.org/10.3390/polym10121365

[8] A. Thamer, H. Y. Ali, J. J. Najwa, TGA, DSC, DTG Properties of Epoxy Polymer Nanocomposites by Adding Hexagonal Boron Nitride Nanoparticles, ARPN J of Engg and Appl. Sci. (14) 2, (2019), 567-574, https://doi.org/10.36478/jeasci.2019.567.574

[9] M. J. Zayza, N. J. Jubier, A. Subhi, Thermal Oxidative of UP/MgO Nanocomposites Using TGA and DSC Analysis, World Journal of Research and Review (WJRR), ISSN: 2455-3956, 10, 2, (2020) 18-24.

[10] C. Vivekanandhan, P. S. Sampath, K. P. Sathish, B. Mylsamy, A. Karthik, Characterization on thermal properties of glass fiber and Kevlar fiber with modified epoxy hybrid composites, J of Matls. and Res. Tech. 9,3 (2020), 3158-3167. https://doi.org/10.1016/j.jmrt.2020.01.061

[11] M. Ramesh, N. R. Lakshmi, N. Srinivasan, D. V. Kumar, D. Balaji, Influence of filler material on properties of fiber reinforced polymer composites: A review, e-Polymers 22 (2022), 898-916. https://doi.org/10.1515/epoly-2022-0080

[12] T. P. Sathishkumar, S. Satheeshkumar, J. Naveen, Glass fiber-reinforced polymer Composites: A review, J of Rei. Plts. and Comp., SAGE Publications, 33 (2014) 1258-1275. https://doi.org/10.1177/0731684414530790

[13] C. M. Ramesh, Experimental Study on Optimization of Thermal Properties of Natural Fibre Reinforcement Polymer Composites, , Open Access Lib. J., 5, 2333-9721, (2018), 01-14.

[14] R. Ambigai, S. Prabhu, Analysis on mechanical and thermal properties of glass carbon/epoxy-based hybrid composites, IOP Conf. Series: Mtls. Sci. and Engg. 402 (2018) 012136, 01-12. https://doi.org/10.1088/1757-899X/402/1/012136

[15] R. Ganesh, P. Anand, D. Wubishet, Experimental Investigation on Thermal Behaviours of Nanosilicon Carbide/Kenaf/Polymer Composite, J of Nano. 3906336, (2022), 01-09. https://doi.org/10.1155/2022/3906336

[16] K. M. Babu, M. Mettilda, Studies on Mechanical, Thermal, and Morphological Properties of Glass Fibre Reinforced Polyoxymethylene Nanocomposite, J of App. Chem., 82618, (2014), 01-08. https://doi.org/10.1155/2014/782618

[17] L. Bhanuprakash, S. Varghese, S. K. Singh, Glass Fibre Reinforced Epoxy Composites Modified with Graphene Nanofillers: Electrical Characterization, Hind. J. of Nano. (2022), 01-08. https://doi.org/10.1155/2022/4611251

[18] M. Farzana, Md. H. Marjanul, S. M. Sonali, M. Z. I. Mollah, Md. A. Amin, S. A. Khan, I. Farjana, R. A. Khan, Thermo-Mechanical Properties and Applications of Glass Fiber Reinforced Polymer Composites, Mod. Copts. in Mat. Sci., RIS publishers, 2692-5397 (2023), 01-09.

[19] B. Barbara, V. Katarina, V. Marko, M. Barbara, Thermal properties of polymer-matrix composites reinforced with E-glass fibers, J of Microele., Eltr. Comp. and Mtls., 45,3 (2015), 216 - 221.

[20] S. Galal, D. I. Chukov, V. V. Tcherdyntsev, V. G. Torokhov, D. Z. Dmitry, Effect of Glass Fibers Thermal Treatment on the Mechanical and Thermal Behavior of Polysulfone Based Composites, 12(04), 902 Polymers 2020, 01-11. https://doi.org/10.3390/polym12040902

[21] N. I. N. Haris, R. A. Ilyas, Md. Z. Hassan, S. M. Sapuan, A. Afdzaluddin, K. R. Jamaludin, S. A. Zaki, F. Ramlie, Dynamic Mechanical Properties and Thermal Properties of Longitudinal

Basalt/Woven Glass Fiber Reinforced Unsaturated Polyester Hybrid Composites, Polymers 3343, (2021), 01-14. https://doi.org/10.3390/polym13193343

[22] T. Klaser, L. Balen, Ž. Skoko, L. Pavic, Š. Ana, Polylactic Acid-Glass Fiber Composites: Structural, Thermal, and Electrical Properties, Polymers 14(19), 4012 (2022), 01-11. https://doi.org/10.3390/polym14194012

[23] N. Siti, A. Syafiuddin, A. B. Hong Kueh, A. Maddu, Physical, thermal, and mechanical properties of polypropylene composites filled with rattan nanoparticles, J of App. Res. and Techn., 15 (2017) 386-395. https://doi.org/10.1016/j.jart.2017.03.008

[24] G. B. Vaggar, S. C. Kamate, S. L. Nadaf, A study on thermal conductivity and thermogravimetric analysis of glass fiber epoxy resin composites modified with silicon carbide and copper nanoparticles. Mtls. Today: Prodgs., 66 (2022), 2308-2314. https://doi.org/10.1016/j.matpr.2022.06.230

Materials Joining and Manufacturing Processes: MJMP 2025   Materials Research Forum LLC
Materials Research Proceedings 55 (2025) 72-78   https://doi.org/10.21741/9781644903612-12

Review on additive manufacturing at the forefront: Exploring recent developments and industry applications

Gurushanth B. Vaggar[1,a*], Elvin Chris Dsouza[1,b], Shyam[1,c],
Ananthesh D. Kamath[1,d]

[1]Department of Mechanical Engineering Alva's Institute of Engineering and Technology,
Moodbidri 574225, India

Affiliated to Visvesvaraya Technological University, Belagavi, India

[a]gvaggar7@aiet.org.in, [b]elvindsouza2004@gmail.com, [c]bhatshyam757@gmail.com,
[d]mech01anantheshkamath@gmail.com

Keywords: Additive Manufacturing, Materials Science, 3D Printing, Statistical Analysis, Industry Applications

Abstract. Additive manufacturing (AM) has experienced significant growth in academia and industry, driven by its ability to produce complex geometries. This technology continues to evolve, addressing initial limitations and expanding its use across various industries. Public interest in additive manufacturing is rising due to its diverse opportunities and applications. This article outlines the fundamentals of the technology, emphasizing key benefits and drawbacks. AM offers design flexibility, mass customization, waste reduction, complex structure production, and rapid prototyping. A detailed review was conducted on key 3D printing methods, materials, and their advancements, focusing on groundbreaking applications in biomedical, aerospace, construction, and protective structures. The integration of engineering design with AM has progressed more slowly than expected. This paper highlights key opportunities, limitations, and economic factors in Design for Additive Manufacturing. It examines challenges related to designing and redesigning for both direct and indirect AM production methods.

Introduction

Additive Manufacturing, more popularly also known as 3D printing, has exploded into tremendous growth since the concept started in the last portion of the 20th century. It was utilized almost as a niche solution. Primarily, it operated merely as a rapid prototyping tool within specific markets in aerospace and automotive industries to bring quicker design iterations for competitive purposes. However, over the past few decades, AM has expanded significantly, transforming into a widespread manufacturing method employed across many sectors, including healthcare, consumer goods, construction, and even food production [1]. As of 2021, the global market has reached approximately $16.4 billion in AM. The ongoing market analysts predict a lot of growth in this space and estimate that the market will reach approximately $76.2 billion by 2030. This growth rate is pretty impressive with a compound annual growth rate (CAGR) of 17.5%, which highlights the growing importance of AM in the overall manufacturing landscape [2][3][4].

Some major enabling factors behind the widespread application of Additive Manufacturing include significant breakthroughs in material science that have made a wide range of applications feasible. Today, diverse polymers, high-performance metals, ceramics, and even bio-materials are available for 3D printing to be used for various applications such as tissue engineering in the medical field. All these have enabled manufacturers to tailor material properties to achieve specific performance criteria required by different industries. Furthermore, AM technologies are also becoming even more accurate and precise, which has facilitated the generation of geometries and designs which were quite impossible with traditional manufacturing methodologies. The Selective

Content from this work may be used under the terms of the Creative Commons Attribution 3.0 license. Any further distribution of this work must maintain attribution to the author(s) and the title of the work, journal citation and DOI. Published under license by Materials Research Forum LLC.

Materials Joining and Manufacturing Processes: MJMP 2025   Materials Research Forum LLC
Materials Research Proceedings 55 (2025) 72-78   https://doi.org/10.21741/9781644903612-12

Laser Sintering (SLS), Fused Deposition Modeling (FDM) and Electronic Beam Melting (EBM) methods all improved the capability of AM for the manufacture of parts that could have complex geometries and very fine detail along with structural integrity [5][6].

Another critical driver, that is increasingly drawing people's attention toward AM, is sustainability. Because most of the traditional manufacturing processes waste a lot and consume lots of energy, it is widely known that AM does not waste material at all. How an object is added layer after layer reduces the amount of waste, besides providing better usage of resources, and, with the mounting emphasis in the sectors towards sustainable practices [7].

This paper will draw upon statistical and analytical data in trying to explore recent technological developments in Additive Manufacturing, as well as the multiple industrial applications. It will try to provide a quantitative overview of the impact of AM upon different sectors, inquiring into how this new manufacturing method modifies the design and nature of production and manufacturing processes (Figure 1). These developments demonstrate how (AM) not only offers operational benefits but also has the potential to revolutionize traditional manufacturing paradigms, paving the way for a more sustainable industrial future [8].

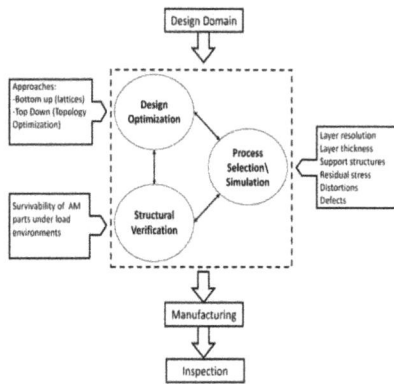

Figure 1. *Overview of AM engineering cycle [8].*

Materials Science in Additive Manufacturing: Progressions in Additive Manufacturing Materials

The global AM materials market, valued at $1.6 billion in 2022, Table 1, demonstrates diverse material utilization across industries. Metals dominate due to their critical role in the aerospace and automotive sectors, followed by polymers and ceramics. Biomaterials, primarily used in healthcare, are experiencing rapid growth [9].

Table 1. *Global AM Material Market (2022) [9].*

Material	Market Share (%)	Growth Rate (CAGR)	Key Applications
Metals	43	15%	Aerospace, Automotive
Polymers	35	12%	Consumer Goods, Prototyping
Ceramics	12	9%	Medical Implants, Electronics
Bio Materials	10	22%	Healthcare, Bioprinting

Materials Joining and Manufacturing Processes: MJMP 2025   Materials Research Forum LLC
Materials Research Proceedings 55 (2025) 72-78   https://doi.org/10.21741/9781644903612-12

Comparative Material Properties

AM materials have a large variability in their mechanical properties, so it is important to understand when the material is applied to a different context. For example, tensile strength refers to the resistance of a material to stretch or pull until breaking, and thermal stability indicates a material's resistance to changing properties at high temperatures. In this regard, Table 2 presents a comparison. Comparison of the differences between these materials makes selection easier in line with the requirement for manufacture. AM materials vary considerably in mechanical performance. 3D printing materials have highly varied mechanical properties. Table 2 presents a comparison of tensile strength and thermal stability for commonly used materials.

***Table 2**. Mechanical Properties of AM Materials [2].*

Material	Tensile Strength (MPa)	Thermal Stability (°C)	Applications
PLA (Polymer)	40-60	200-220	Consumer goods, prototypes
Ti-6Al-4V (Metal)	900-1200	800-1000	Aerospace, automotive
Alumina (Ceramic)	300-600	>1200	Electronics, implants
PCL (Biomaterial)	10-15	50-60	Tissue engineering, prosthetics

Technological Advancements in Additive Manufacturing: ***Growth in Printing Techniques***

Leading-edge AM techniques such as Directed Energy Deposition (DED) and Powder Bed Fusion (PBF) (Figure 2a) cumulatively account for roughly 60% of the overall market in the industrial segment. DED and PBF are relatively new processes that provide high definition and exceptional cohesiveness in the resulting component. DED is one of those favorable methods, pronounced as the most diversified technique allowing large-scale metal piece manufacturing. The melting and deposition in a controlled manner allow repairing and modifying former parts along with making new ones. One area where this is most commonly utilized is in the aerospace industry, where it manufactures critical components that demand strict performance and reliability standards. Laser Metal Deposition (LMD) is an AM technique that uses DED. In LMD, a high-powered laser melts metal powder or wire, which is then precisely deposited layer by layer to build a 3D object. This method enables high precision and complex geometries in manufacturing.

PBF, on the other hand, is much famed for making very complicated and complex geometries and shapes. Selective Laser Melting (SLM) is a PBF (Figure 2d), AM technique. In SLM, a laser selectively melts metal powder particles layer by layer on a bed, gradually building a 3D object with high precision and detail. This application involves the melting of powders layer by layer, selectively and therefore suits industries such as hospitals, diagnostics, pharmaceuticals, medical devices, and telemedicine sectors. Within health care, PBF technology is used for the fabrication of specific implants, prosthetics, and surgical tools with high precision and customization for the needs of each patient. Both DED and PBF, however, remain the path-generators for innovation in their respective industries and continue to inspire change in the functionalities made possible by additive manufacturing techniques (Figure 2b, 2c) in producing products for various uses.

Integration of Automation

By 2025, 85% of AM processes will be anticipated to incorporate automation and real-time monitoring systems, facilitating a 30% reduction in error rates. These technological advancements not only enhance the scalability of operations but also significantly improve their reliability. The market share and growth rate of AM are given in Table 3.

Materials Joining and Manufacturing Processes: MJMP 2025   Materials Research Forum LLC
Materials Research Proceedings 55 (2025) 72-78   https://doi.org/10.21741/9781644903612-12

Adoption Across Industries

Table 3. AM Market Distribution by Industry (2022) [5].

Industry	Market Share (2022)	Growth Rate (CAGR)	Key Applications
Aerospace	35%	16%	Lightweight parts, fuel nozzles
Healthcare	28%	22%	Prosthetics, dental implants, bioprinting
Automotive	18%	14%	Prototyping, tooling, limited-run parts
Consumer Electronics	12%	10%	Custom casings, heat sinks

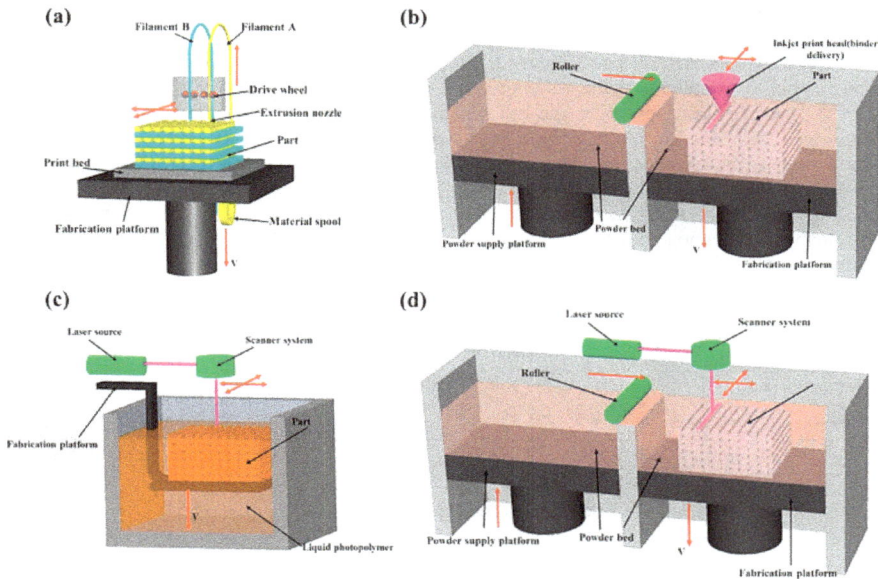

Figure 2. Schematic diagrams of four main methods of additive manufacturing: (a) Fused deposition modelling; (b) Inkjet printing; (c) Stereolithography; (d) Powder bed fusion [1].

Economic and Environmental Impacts: *Cost Savings*

Built-up AM has tremendous benefits for traditional sub-constructs that are detached through its power to reduce waste by as much as 90% material: it has benefited tremendously from this. Waste is said to be beneficial not only from the perspective of resource use but also from the perspective of cost (Figure 3). Industries like aerospace, which prioritize the precision and efficiency of their materials, reap significant benefits from AM technology. This is a significant reduction in costs due to the reduction of unusable off-cuts, as well as increased efficiency in using raw materials. It would result in the health sector by making medical solutions cheaper and more accessible, as manufacturers would be able to allocate the resources more strategically [10].

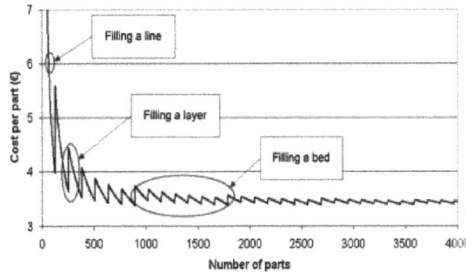

Figure 3. Cost per part vs. the number of parts produced estimated [11].

Environmental Benefits

Indeed, the environmental impacts of AM are very promising. Localized manufacturing is contributing significantly to reduced supply chain emissions. By reducing the distance that materials travel, AM has been shown to result in up to a 25% reduction in emissions, thus contributing to a smaller overall carbon footprint.

AM is especially relevant in the era of today where sustainability continues to grow in importance. Closing the loop recycling systems into AM processes complements further improvements in dimensions. They also keep resources optimally and accountably used throughout their entire lifecycle by promoting the circular economy approach, which involves reusing materials and minimizing waste (Table 4). All these will make AM a frontrunner in sustainable production practices, well-gifted for appealing to environmentally conscious organizations [12].

Table 4. Environmental Benefits of AM [12].

Impact Area	Reduction Achieved	Applications
Material Waste	90%	Aerospace, Prototyping
Carbon Emissions	25%	Localized Manufacturing
Energy Consumption (per unit)	20-30%	Metal and Polymer Printing

Challenges in Additive Manufacturing: *Scalability Issues.*

Compared to subtractive manufacturing, AM significantly reduces material waste by up to 90%. This reduction translates into substantial cost savings, particularly in the aerospace and healthcare industries. Although technological advancements have been made, current AM systems are still 30-40% slower than traditional methods for large-scale production (Table 5). This discrepancy impedes adoption in high-volume industries such as automotive [13].

Healthcare Industry

That is, over 70% of items produced through additive manufacturing face a multi-stage certification process in the healthcare area. This certification is so extensive because of the stringent safety and efficacy requirements for medical use. It is extensive because testing and verification must be done at various stages in the product's development process, which slows market entry. Such delays stifle innovation and restrict patient access to newer devices that could be beneficial in treatment. Hence, the complicated regulatory framework for AM in healthcare stands as an incredible obstacle to fast growth in many promising areas [14].

Materials Joining and Manufacturing Processes: MJMP 2025
Materials Research Proceedings 55 (2025) 72-78

Materials Research Forum LLC
https://doi.org/10.21741/9781644903612-12

Limited Material Choices

While a considerable number of materials have emerged for AM, it should be borne in mind that only a small amount, approximately 10-15%, of high-performance alloys and ceramics have been used in AM systems. This places a limit on the possible applications and performance capability of AM technologies since most of the high-performance materials cannot find applications within these existing frameworks and processes. This further creates an urgent need for more research and development for opening high-performance materials to AM so that optimum versatility can be achieved with innovation across industries [15].

Table 5. Challenges in AM Adoption [16].

Challenge	Metric	Industry Affected
Production Speed	30-40% slower than traditional	Automotive, Consumer Goods
Certification Time	1-2 years per component	Aerospace, Healthcare
Material Compatibility	10-15%	Multi-material applications

Future Research Direction.

- **Material Innovation:** Expand compatibility for high-performance alloys and biocompatible materials.
- **Process Automation:** Develop AI-based systems to improve scalability and precision.
- **Sustainability:** Design closed-loop systems for material recycling and waste reduction.

Conclusions

- Thereby, AM is going up as a transformational technology in production. Analytical studies show the advantages associated with it in customizing, reducing environmental influence, and optimizing resource usage. Issues relating to scalability, diversity of materials, and being proven through regulations, however, still need to be addressed in the years to harness the full potential of this new technology.
- The sustainability benefits, efficient resource use, and versatility in instruments make it one of the key driving forces of innovation in manufacturing today. AM application faces challenges concerning its widespread applicability such as scalability barriers, material limitations, as well as a rather dense and intricate framework of regulatory requirements. With ongoing technological advances as well as academic explorations and research, these problems can ultimately be solved in the long run for AM to prove its full potential for being an important asset in industrial production.

References

[1] T. Ngo, A. Kashania, G. Imbalzanoa, K. T. Q. Nguyena, D. Huib, Additive manufacturing (3D printing): A review of materials, methods, applications, and challenges, Composites Part B, 143, (2018) 172-196. https://doi.org/10.1016/j.compositesb.2018.02.012

[2] M. Fateri, A. Gebhardt, Introduction to Additive Manufacturing, Springer, Cham., 233 (2020) 01-22. https://doi.org/10.1007/978-3-030-58960-8_1

[3] A. Kosaraju, Material Challenges in Additive Manufacturing, Mtls. Sci. J, 126 (2019) 365-375.

[4] T. Shah, AI Integration in Additive Manufacturing, Ind. AI J, 22, (2021) 50-65.

Materials Joining and Manufacturing Processes: MJMP 2025   Materials Research Forum LLC
Materials Research Proceedings 55 (2025) 72-78   https://doi.org/10.21741/9781644903612-12

[5] B. Ahuja, Additive Manufacturing Adoption Across Industries, J. of Mfg. Proc., 62, (2022) 80-91.

[6] C. Roach and J. Gardner, Economic Impact of Additive Manufacturing, Mfg. Eco. Review, 54, (2020) 12-23.

[7] X. Wang, M. Jiang, Z. Zhou, J. Gou, D. Hui, 3D printing of polymer matrix composites: A review and prospective, Comp. Part B: Engineering. 100 (2017) 442-458. https://doi.org/10.1016/j.compositesb.2016.11.034

[8] B. Barroqueiro, A. Andrade-Campos, R. A. F. Valente, V. Neto, Metal Additive Manufacturing Cycle in Aerospace Industry: A Comprehensive Review. Journal of Manufacturing and Materials Processing, 3(3), 52 (2019) 01-21. https://doi.org/10.3390/jmmp3030052

[9] M. G. Pérez, D. Carou, E. M. Rubioa, R. Teti, Current advances in additive manufacturing technologies, Procedia CIRP, 91 (2020) 439-444. https://doi.org/10.1016/j.procir.2020.05.076

[10] I. Gibson, Sustainability Metrics in Additive Manufacturing, Additive Manufacturing Handbook, 2015. https://doi.org/10.1007/978-1-4939-2113-3

[11] M. K. Thompson, G. Moroni, T. Vaneker, G. Fadel, R. I. Campbell, I. Gibson, A. Bernard, J. Schulz, P. Graf, B. Ahuja, F. Martina, Design for Additive Manufacturing: Trends, opportunities, considerations, and constraints, CIRP Annals - Mfg. Tech., 65 (2016) 737-760. https://doi.org/10.1016/j.cirp.2016.05.004

[12] A. M. Beese, Environmental Benefits of Additive Manufacturing, Advanced Manufacturing Handbook, 2018.

[13] T. Mukherjee, W. Zhang, T. DebRoy, An improved prediction of residual stresses and distortion in additive manufacturing, Comp. Mtls. Sci., 126, (2017) 360-372. https://doi.org/10.1016/j.commatsci.2016.10.003

[14] A. Lundbeck, Regulatory Barriers in Additive Manufacturing, Aero. Mtls. J, 14, (2020), 105-117.

[15] M. Hotza, Material Limitations in Additive Manufacturing, Adv. Mtls. Re., 45, (2016) 190-201.

[16] J. Kang, Multistage Certification in Additive Manufacturing, J. of Indl. Engg, 25, (2021) 145-159.

Materials Joining and Manufacturing Processes: MJMP 2025
Materials Research Proceedings 55 (2025) 79-84

Materials Research Forum LLC
https://doi.org/10.21741/9781644903612-13

Implosive reactive armour to enhance explosive welding on hard surfaces in modern warfare

A. Gyanesh Kumar Rao[1a] *, Satyanarayan[2b], Suresh Ganesh Kulkarni[3c]

[1]Gyanadraksha Wydhumraketustra Subrahmkr Private Limited, Cabin 2, Rubi Tower 2, Rungta Educational Campus, kurud, Durg, 490024, Chhattisgarh, India

[2] Department of Mechanical Engineering, Alva's Institute of Engineering and Technology Karnataka, Moodbidri, Mangalore 574225 and affiliated to Visvesvaraya Technological University, Belagavi, Karnataka, India

[3]Department of Explosives and Applied Chemistry, Defence Institute of Advanced Technology (Deemed University), Girinagar, Pune, 411025, Maharashtra, India

[a]gwsdefence@gmail.com, [b]satyan.nitk@gmail.com, [c]sgk_iat1@rediffmail.com

Keywords: Implosive Reactive Armour, Modern Warfare, Joggle Joining

Abstract. The study focuses on the role Implosive Reactive Armour (IRA) in enhancing explosive welding techniques for hard surfaces in modern warfare. IRA is a tertiary layer of protection that fills the gap between traditional explosive reactive armour (ERA) and Vaporizing Reactive Armour (VRA). IRA is designed to protect against shaped-charge warheads and improve vehicle survivability in combat. The IRA's joggle interlocking S profile of HRC grade steel plates features plasticized nitrocellulose blend with TNT explosive cladding and graphene-coated sisal fiber over rubberized NIJ 3+ composite armour backing of Ultra-High-Molecular-Weight Polyethylene (UHMWPE). The IRA significantly enhances protection, exhibiting varying levels of effectiveness depending on the type of threat, design of the reactive elements, and placement on the vehicle. The explosive welding process for joining HRC-grade steel plates requires careful control of the explosive energy, as these steels are hard and prone to cracking under excessive force. The binder material's properties, such as its ability to cushion and regulate detonation, play a significant role in the quality of the final weld. The findings suggest that IRA remains a valuable defence tool but must be adapted or combined with other technologies in quad armour to ensure maximum effectiveness against future threats. Continuous innovation is required to maintain IRA's relevance in modern warfare to mitigate beyond Stanag level 6 fragmentation round threats varying from ground to space debris by infusing explosive welding techniques in it. IRA side walls profile in joggle or corrugated roofing sheet container like profile and fuel tanks providing upper hand in stealth operations with no visible design changes in outer body of the heavy vehicles and aircraft hangers. The goal of this research is to assess the effectiveness, limitations and future potential of IRA in modern combat scenarios with a focus on its ability to counter act on evolving threats being an add on protection between a quad armour layer of metal matrix composite involving explosive welding

Introduction

Modern warfare has undergone considerable change. The use of drones in recent warfare has challenged the concept of traditional warfare. Protection of assets needs a complete relook as identification and destruction have become much easy. IRA side walls profile in joggle or corrugated roofing sheet container like profile & fuel tanks providing upper hand in stealth operations with no visible design changes in outer body of the heavy vehicles. It has got pre oxidized acrylic fibers-based carbon nano tubes doped with shear thickening fluids made up of

Content from this work may be used under the terms of the Creative Commons Attribution 3.0 license. Any further distribution of this work must maintain attribution to the author(s) and the title of the work, journal citation and DOI. Published under license by Materials Research Forum LLC.

aerosol and zeolites for better performance as heat shield under high temp shock impacts via cladding to function as radar obscurant cloak while being fused between VRA and ERA.

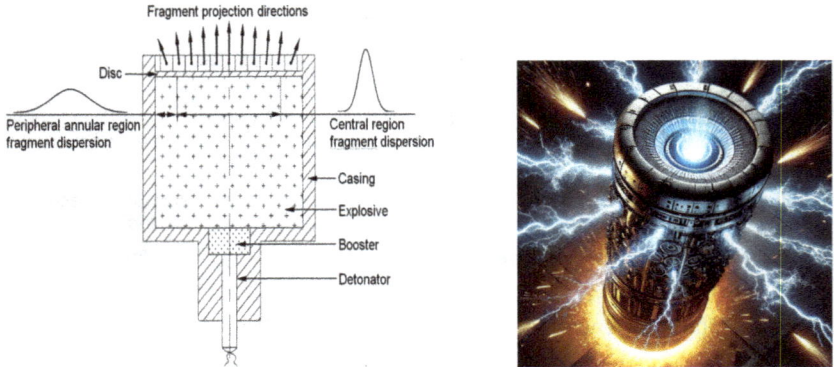

Fig.1 *(a) ERA plate style discharge system [1-3]   (b) VRA trying to melt 120 mm mortar casing*

Here in Fig.1a we can clearly observe ERA uses the concept of metallic flyer plate to be ejected post incoming tip of high-speed projectile leads to tou   ch critical army assets like battle tanks in majority of tanks in the form of tiles whereas very less known VRA (Fig. 1b) was initially used in trials decades ago just to burn the outer cartridge of incoming high-speed projectile with n heavy discharge of both current and voltage from series and parallel connection of batteries to function as bulky super capacitor. This later was avoided due to high recharging duration in remote areas and in mean time enemy weapon could have proceed further with heavy damages on tanks either via armour piercing rounds or anti-tank guided missiles.

Also, we are familiar that explosive welding eliminating crevice corrosion issues inherent in mechanical joints like rivets and bolts, offering a solution to galvanic corrosion challenges. As the shipbuilding industry shifts towards lightweight designs for more efficient ships, explosive welding remains integral to constructing vessels that meet modern design and efficiency standards will be rise in demands for small suicide boats [4,5]. Mass reduction is a critical strategy in automotive engineering to minimize fuel consumption. The joining of different materials, such as aluminum and copper, is frequently employed in electrical wiring systems, like bus bars and connectors. These connections are commonly utilized in conjunction with batteries for electric vehicles. This hybridization ensures effective electrical conductivity while also lowering weight and cost [6,7].

Methodology

The Table 1 structural layer shall follow the mirrored pattern across 4mm HRC steel to initiate explosive welding via implosion behaviour post the recoil of ERA panels or any catastrophic shock from either side of the 6 mm fusion panels. Sandwiched between VRA and IRA.

Now as such to further enhance its experimental results on fusing two dis-similar grades of material the composite NIJ 3+ backing all in dimensions of 500 x 500 [mm] were taken for initial trials that was later modified into S shape joggle structure with respective coatings and explosive fillers in form of cladding sheet of approximately 2.5 [mm] thickness as shown in Fig.2. Arranged further within Fig. 3 by the nomenclature of air gap whose structure is described within Table 1 . Its properties have shown significant effects both in simulation and field trials to resist beyond shock impacts of 3 to 4 [mm] fragments coming at the standalone speed of 3400 [m/s]. Here HNS

Materials Joining and Manufacturing Processes: MJMP 2025
Materials Research Proceedings 55 (2025) 79-84

Materials Research Forum LLC
https://doi.org/10.21741/9781644903612-13

hardness is of around 400 whereas Armox advance and 650 T harness is between 620 and 650. The composition of Armox is provided in Table 2.

Table 1 IRA panel structure inside air gap nomenclature

Graphene coated 4 bio armor net having heavy Lignin extract from peroxy-formic acid with sisal	6mm HRC steel (strike face)	Powder Coating primary explosive- Lead Dinitrores -corniate (10%)	Plastic based cladding secondary explosive blend like thermocol sheet type structure of 5:3 RDX & HTBP (80%)	Powder Coating primary explosive- Lead Dinitrores -corniate (10%)	4mm HRC steel (fusion face)	Powder Coating primary explosive- Lead Dinitrores -corniate (10%)	Plastic based cladding secondary explosive blend like thermocol sheet type structure of 5:3 RDX & HTBP (80%)	Powder Coating primary explosive- Lead Dinitrores -corniate (10%)	6mm HRC steel (strike face)	Graphene coated 4 bio armor net having heavy Lignin extract from peroxy-formic acid with sisal

(a) Graphene coated Bio armour net
(b) Sisal based lignin
(c) 4 mm & 6 mm Armox Advance ,620T
(d) HB26- PVB layered acrylic sheets & glass
(e) composite matrix sheets of HNS,AL7075,SiC,B4C, WC, Ti6Al4V, KFRP, UHMWPE

Fig.2 Composite materials involved in development of IRA and its backing

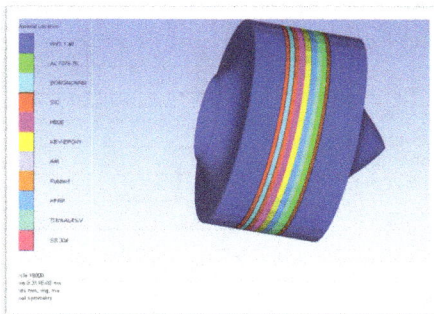

Fig.3 Fragmentation impact test in LS Dyna, Ansys 2022 R1

Table 2: Armox composition

Chemical Composition									
Heat No	**C**	**Si**	**Mn**	**P**	**S**	**Cr**	**Ni**	**Mo**	**Cu**	**B**
141789	.43	.15	.29	.005	.001	.20	1.80	.146	.01	.001

Materials Joining and Manufacturing Processes: MJMP 2025   Materials Research Forum LLC
Materials Research Proceedings 55 (2025) 79-84   https://doi.org/10.21741/9781644903612-13

The interaction between these layers is optimized through precision explosive welding techniques caused by implosive recoil energy and pyrotechnics of shorting Nickle Cadmium battery with a wireless switch, ensuring seamless energy transfer and synchronization across the structure to fuse [8]. Strong bonds can be produced by the explosive welding process and usually the weld interface has a characteristic wavy form [9].

Empirical relations for explosive driven plates

C= Total explosive mass per unit area, M= Total mass per unit area, E= Gurney Energy
$\sqrt{2E}$ = Gurney Characteristic Velocity,

$$V = \sqrt{2E} \left[\frac{\left(1 + \frac{2M}{C}\right)^3 + 1}{6\left(1 + \frac{M}{C}\right)} + \frac{M}{C} \right]^{\frac{-1}{2}}$$

Results
In this modification of IRA in real time application has been tried with roll up and roll down joggle sheets to give beyond Fragment Sharpenels Protection (FSP) of Stanag 6 (Fig 4a and Fig 4b). Hence these joggle HRC steel structured ballistic materials to redirect and infuse blast energy in to weld joint, thereby harnessing the energy of destructive force and preventing an escalation of sympathetic explosion hazards in closely parked or moving vehicle convoys and aircrafts. So that need for recovery vehicles and cranes can also be minimized during war times and burden of overhauling can be reduces at base repair depots. Table 3 provides information on the Fragment Shrapnel Protection comparison with IRA.

Table 3*. Fragment Shrapnel Protection comparison with IRA*

Velocity along the length of the warhead starting from the point of detonation	Velocity by high speed Photographic technique (m/s) as per TBRL Report	Velocity calculated through the computer code (m/s)
Our Implosive reactive armour can stop fragments upto 3400 (m/s) considering factor of safety for bigger fragments rounds- **Ref Fig. 5**	2727	2465.935
	2721	2461.884
	2616	2449.604
	2613	2428.703
	2611	2398.452
	2513	2357.639
	2332	2304.269
	2331	2234.868
	2323	2142.602
	2177	2009.946
Avg. ejection velocities of the fragments	1977	1657.6

Materials Joining and Manufacturing Processes: MJMP 2025
Materials Research Proceedings 55 (2025) 79-84

Materials Research Forum LLC
https://doi.org/10.21741/9781644903612-13

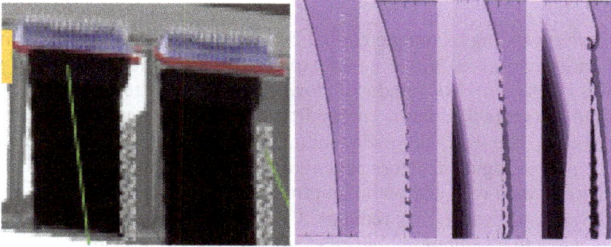

(a)S- shaped joggle in white   (b) Fused IRA 4mm & 6mm

Fig.4 Fused plates in air gap region of composite panels of armour joggle structure shutter

So, ultimately post VRA does its job for melting combustible cartridges, fragmentation rounds will further lose its lethality with ERA flyer plate. Then finally, due to implosion by recoil of ERA explosive welding will give an add on layer of armour protection.

Discussions

It has got significant application in modern warfare to enhance infantry protected vehicles and other military, navy, air force, spacecraft, satellite assets against hyper velocity impacts on these military assets. IRA embedded within the structural framework, designed to activate upon explosive impact to redirect and contain blast energy, thereby reducing the risk of structural damage. Ultimately, it reduces the jet velocity of 7 to 14 km/s to 2 Km/s that is been protected by IRA operating at factor of safety to operate till giving protection up to 3.4km/s (Fig 5a and 5b).

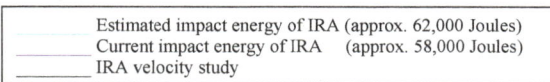

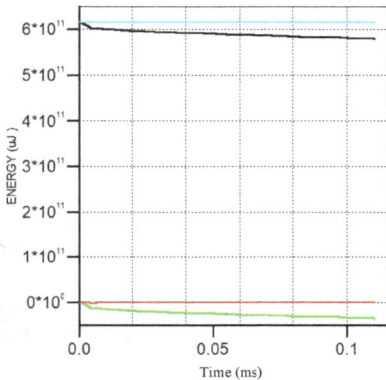

Fig.5 (a) *Energy comparison in IRA*   (b) *Velocity comparison in IRA*

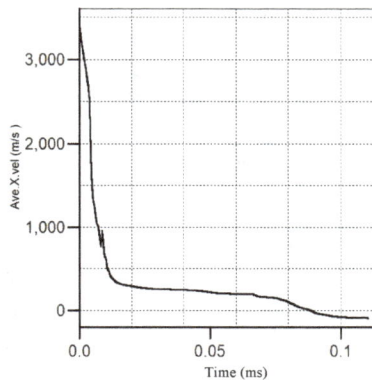

Summary

- It is explosive welding technique caused by recoil leading to implosion in the implosive reactive armor ignites its primer based explosive cladding layer precisely upon impact to fuse a weld between HRC grade rolldown shutter type steel plates, which effectively absorbs and disperses the kinetic energy from the IRA's primer charged projectile.
- Simultaneously, the ballistic-grade netting on the outer layer captures any residual fragments, preventing debris from penetrating the hangar and ensuring that any secondary damage is contained with add on bullet proof glasses.
- The findings suggest that IRA remains a valuable defence tool but must be adapted or combined with other technologies in quad armour to ensure maximum effectiveness against future threats.

Abbrevation

SiC- Silicon Carbide, B4C- Boron Carbide, WC- Tungsten Carbide, KFRP- Kevlar Fibre Reinforced Polymer

References

[1] K .D Dhote, KPS Murthy, K.M. Rajan, M. M. Sucheendran "Statistics of fragment dispersion by explosion in fragment generator warhead. Central European", J. Energ. Mater. 13(1) (2016)183-197. https://doi.org/10.22211/cejem/64971

[2] M.R Driels Weaponeering Conventional Weapon System Effectiveness. Reston, Virginia, USA:American Institute of Aeronautics and Astronautics.(2004)

[3] M. Hartmann M Component Kill Criteria a Literature Review. FOI, Swedish Defence Research, (2009) Agency.Available at: www.foi.se/ReportFiles/foir_2829.pdf (assessed 20 January 2016).

[4] P. Corigliano, V. Crupi, E. Guglielmino, A.M. Sili, Full-field analysis of AL/FE explosive welded joints for shipbuilding applications, Mar. Struct. 57 (2018) 207-218. https://doi.org/10.1016/j.marstruc.2017.10.004

[5] Y. Kaya, Microstructural, mechanical and corrosion investigations of ship steel-aluminum bimetal composites produced by explosive welding, Metals 8 (7) (2018) 544. https://doi.org/10.3390/met8070544

[6] J.P. Bergmann, F. Petzoldt, R. Schürer, S. Schneider, Solid-state welding of aluminum to copper-case studies, Weld. World 57 (2013) 541-550. https://doi.org/10.1007/s40194-013-0049-z

[7] Y. Kaya, Investigation of copper-aluminium composite materials produced by explosive welding, Metals 8 (10) (2018) 780. https://doi.org/10.3390/met8100780

[8] Rising Phoneix Consultancy LLP, Gyanadraksha Wydhumaketustra Subrahmkr Pvt Ltd. Indian Patent 202421090847. (2024) Wide range aircraft hangar protection system for ballistic and explosive resilience in military and civilian aircraft facilities.

[9] Bahrani A.S., Black T. J. and Crossland Bernard. 1967 The mechanics of wave formation in explosive welding. Proc. R. Soc. Lond. A 296: 123-136 https://doi.org/10.1098/rspa.1967.0010

Materials Joining and Manufacturing Processes: MJMP 2025
Materials Research Proceedings 55 (2025) 85-94

Materials Research Forum LLC
https://doi.org/10.21741/9781644903612-14

Exploring advanced micro-forming methods: Ultrasonic vibration (UV), pulsed current, and laser-assisted approaches

Mihiretu Ganta[1,2,a*], Mahaboob Patel[2], Tewodros Dawit[2]

[1]Faculty of Mechanical Engineering, Opole University of Technology, 45-271 Opole, Poland

[2]Department of Mechanical Engineering, Wolaita Sodo University, Wolaita Sodo, Ethiopia

[a]mihiretu.ganta@student.po.edu.pl

Keywords: Electrically Assisted Forming (EAF), Laser Shock Forming, Micro Forming, Micro-Manufacturing, Ultrasonic Vibration Assisted (UVA) Micro Forming

Abstract. The demand for micro-forming processes has grown due to the increasing need for miniaturization in electronics, medical devices, aerospace, and automotive industries. However, materials with high strength, corrosion resistance, and low weight often lack adequate formability at ambient temperature. Notable examples include titanium alloys, magnesium alloys, and aluminum alloys. Advancements in manufacturing technologies have made micro-forming more viable and cost-effective, leading to a higher demand for smaller and more intricate components. Emerging techniques such as electrically assisted micro-forming, laser shock micro-forming, and ultrasonic vibration-assisted micro-forming have been devised to overcome limitations associated with conventional micro-forming processes, offering enhanced precision, versatility, improved formability, and superior cost-efficiency. This article presents a comprehensive overview of state-of-the-art methods for the aforementioned emerging processes. It identifies existing research gaps, highlights future directions, and examines contact-related challenges in micro/meso-forming processes.

Introduction

Micro-manufacturing refers to producing miniature components with dimensions on the micron or sub-micron scale. It involves using specialized techniques, tools, and processes to fabricate small-scale products. Micro-forming (MF), a subset of micro-manufacturing processes, has recently garnered considerable interest owing to its immense potential in manufacturing small-scale products [1]. The MF process offers several advantages including enhanced productivity, reduced material usage, and versatility in processing various materials. However, since producing micro parts is more challenging and requires more precision, this technique presents several challenges that must be addressed to fully harness its potential [2]. Conventional MF methods (FIG. 1), such as micro-deep drawing (MDD), micro-extrusion, micro-rolling, and micro-stamping methods have been used to produce miniature parts. Even though different researchers tried to improve the micro-forming process of conventional manufacturing methods [3], the efficiency of these techniques has not improved as required. To ensure defect-free products with minimal degradation, it is crucial to optimize both the process parameters and framework of the conventional methods. To overcome the constraints in micro-forming processes, researchers have pioneered innovative approaches, such as laser shock forming (LSF), electric-assisted forming (EAF), and ultrasonic vibration-assisted (UVA) micro-forming. These recent advancements have gained the attention of researchers and industries due to their ability to improve formability, design freedom, better surface finish, energy efficiency, improved tool and die life, lead time reduction, and productivity. EAF is used to improve the formability of difficult materials by applying electrical current to enhance the plasticity and reduce forming forces [4].

Content from this work may be used under the terms of the Creative Commons Attribution 3.0 license. Any further distribution of this work must maintain attribution to the author(s) and the title of the work, journal citation and DOI. Published under license by Materials Research Forum LLC.

Materials Joining and Manufacturing Processes: MJMP 2025   Materials Research Forum LLC
Materials Research Proceedings 55 (2025) 85-94   https://doi.org/10.21741/9781644903612-14

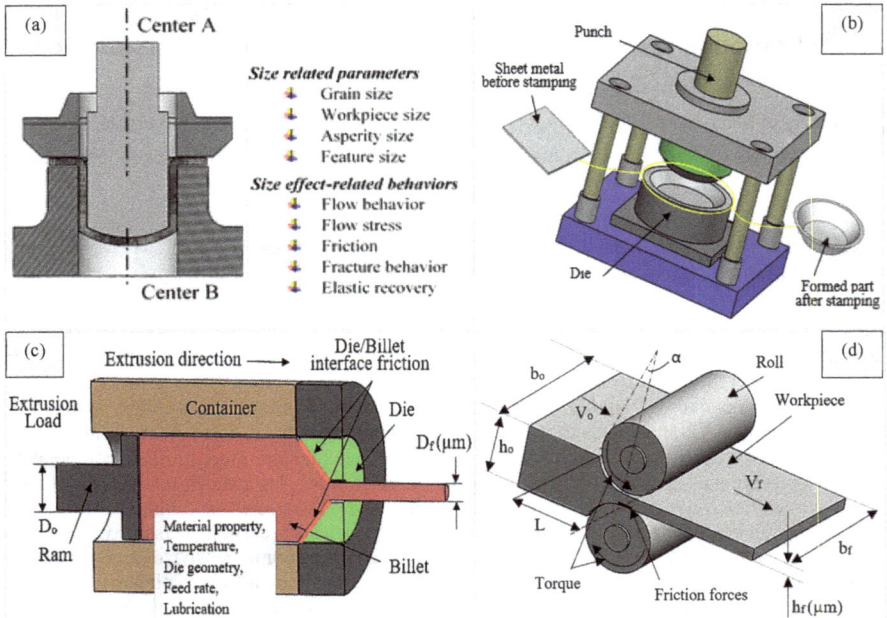

FIG. 1. Conventional micro-forming methods (a) micro-deep drawing, (b) micro-stamping (c) micro-extrusion, and (d) micro-rolling

The effectiveness of EAF has been well established in the realm of reducing forming flow stress, enhancing formability, and minimizing the spring-back effect in various alloys that are typically difficult to form [5]. UVA micro-forming is a method that utilizes high-frequency mechanical vibrations to enhance the formability of materials during micro-forming processes. The application of UV can improve material flow [6], reduce forming forces [7], and enhance the quality of the formed components [8]. Laser shock microforming uses focused laser pulses to create shock waves, enabling precise micro-scale shaping of metals. Owing to its dynamic plastic forming behavior, good filling performance, more uniform material flow, and high forming accuracy were demonstrated by this method [9]. Among the myriad factors that influence the applicability of emerging micro-forming methods, this article offers a succinct overview of the effects of step size, amplitude, frequency (f), lubrication, and other key variables.

Emerging trends in micro-forming

Laser shock micro-forming
Laser shock micro-forming (LSMF) is a high-strain rate technique where intense laser pulses induce rapid, nonlinear deformation, enabling the fabrication of micro-scale components in a short time [10]. Dynamic plastic forming provides high precision, uniform material flow, and superior filling capability. In LSMF, as illustrated in FIG.2, a short-pulse, high-energy laser targets a K9 glass constraint layer. This laser exposure transforms a high-purity aluminum foil into plasma, generating a shock wave that forces the material downward. When the induced pressure surpasses the yield strength, the workpiece undergoes plastic deformation, conforming to the micro-die

Materials Joining and Manufacturing Processes: MJMP 2025
Materials Research Proceedings 55 (2025) 85-94

Materials Research Forum LLC
https://doi.org/10.21741/9781644903612-14

shape [9]. To quantitatively study the influence of laser energy on the forming materials, the pressure generated by the laser is estimated by Fabbro et al. [11], as shown in Eq. 1 and Eq. 2.

$$P_{max} = 0.01 \times \left(\frac{\alpha}{2\alpha+3}\right)^{1/2} \times Z^{1/2} \times I_0^{1/2} \tag{1}$$

$$I_0 = \frac{4E}{\pi D^2} ; \frac{2}{Z} = \frac{1}{Z_{metal}} + \frac{1}{Z_{overlay}} \tag{2}$$

Here, α is a constant, typically set at 0.1, while Z represents the acoustic impedance between the target and the constraint layer. I_0 denotes power density, E corresponds to laser energy, and D refers to the laser spot diameter. Additionally, Z_{metal} and $Z_{overlay}$ indicate the impedance of the target and the constraint layer, respectively. The thermal expansion and contraction of both the mold and sheet are not considered, which may influence forming precision. Furthermore, although laser shock sheet-bulk forming is well-suited for shaping complex sheet components, its dependence on extended process sequences and intricate loading conditions complicates the process design [12]. The effectiveness of LSMF is influenced by laser energy and material temperature. Higher laser energy and elevated temperatures enhance forming depth; however, excessive heat may lead to material ablation. Additionally, annealed workpieces exhibit greater deformation depth compared to unannealed ones due to improved material ductility [13]. Optimizing laser parameters is crucial for achieving uniform micro-die filling. Insufficient energy results in partial mold contact and uneven deformation, while higher energy enhances both forming uniformity and depth [10]. López et al. [14] investigated picosecond LSMF of 316L stainless steel using a 1064 nm solid-state ps-pulsed laser with a 13 ps pulse duration, a repetition rate of 1–20 kHz, and a pulse energy of 115 μJ. While picosecond lasers are generally considered to minimize thermal effects, high repetition rates lead to cumulative heating due to incomplete energy dissipation, which reduces the effectiveness of shock wave-induced stress. The study emphasizes the importance of considering thermal accumulation at high repetition rates to ensure accurate process outcomes.

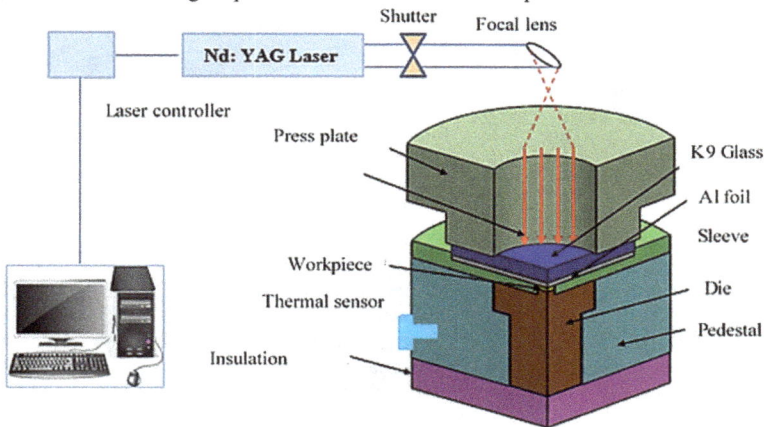

FIG. 2. Schematic diagram of the LSF

In another way, Hou et al. [13] studied the effect of laser loading on grain size, grain size on the depth of forming, and microhardness on AZ31B magnesium alloy. The results show that under the same laser energy loading, the larger the grain size is observed with the more obvious grain size refinement. The forming depth increases with growing grain size, whereas the microhardness

Materials Joining and Manufacturing Processes: MJMP 2025   Materials Research Forum LLC
Materials Research Proceedings 55 (2025) 85-94   https://doi.org/10.21741/9781644903612-14

decreases. Numerical simulation results indicate that under the same laser energy loading, workpiece forming depth decreases with growing die feature groove width. As the grain size of the workpiece increases, the chance of dislocation plugging in the forming process becomes smaller and the dislocation density decreases, so the microhardness of the formed part decreases with the increase of the grain size [13].

Electrically assisted micro-forming

Electrically assisted micro-forming (EAMF) is an emerging metal-forming method that has shown promise in improving a material's formability during deformation. As presented in FIG. 3, the essential components of this setup include a controllable power supply, a DAQ for collecting mechanical data, a thermal camera for capturing thermal information, and insulation [15]. The application of electric current has a significant effect on effectively imparting a localized softening. This can be likened to a form of preheating, albeit with reduced energy losses since only a specific area of the workpiece is being heated. As EAMF is a promising process due to its advantages, such as reducing forming load and increasing formability, the process was successfully employed on explosive welded Ti/Al composite [16], copper sheets micro-rolling [17], micro-compression of Ti alloy [18], AZ31 Mg foils [19], and others.

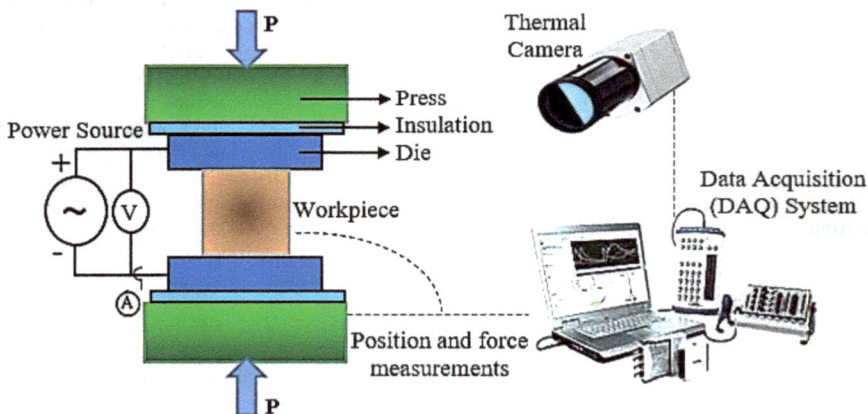

FIG. 3. Schematic of an EAF test setup

To address the limitations of traditional forming processes, Jiang et al. [20] investigated the influence of electric assisted incremental forming process in the production of TC4 titanium alloy. The employed emerging method reduced the need for expensive heating furnaces and reduced the cycle periods. By applying pulse power with a current density range ($4.0 \sim 10.9$ A/mm^2) and a frequency of 140 Hz, the titanium alloy sheet is successfully formed with improved properties. The findings indicate the improvement in material plasticity and effective control of the spring-back effect, resulting in formed parts closely matching the desired geometry. The material's behavior in EAMF is intricately linked to the careful selection of current density and deformation temperature; this is proved by studying the electroplasticity behavior of Ti6554 titanium alloy under different process parameters [21]. Besides the application of reducing forming forces, pulsed current (PC) assisted forming is proposed alternatively in place of the annealing process. Liu et al. [22] suggest using PC instead of elevated temperature and lengthy processing to treat cold-rolled ultra-thin strips of 304 stainless steel. Applying a 25 W pulse current for 5 minutes resulted in complete recrystallization, indicating that PC improves the plastic deformation ability of cold-

Materials Joining and Manufacturing Processes: MJMP 2025   Materials Research Forum LLC
Materials Research Proceedings 55 (2025) 85-94   https://doi.org/10.21741/9781644903612-14

rolled ultra-thin strips by promoting recrystallization and grain refinement [22]. Optimal timing of the electric current, including preheating and total current time, led to a notable reduction in hardness. Erdene-Ochir et al. [23] conclusively demonstrated these by studying the effects of PC on the electrically assisted indentation of SS 304, SS 316, and titanium substrates. Additionally, increasing the amplitude of the electric current further decreased the hardness. To enhance the forming quality of hot-rolled H62 brass with a thickness of 50 μm under high strain rates, Sun et al. [24] employed PC-assisted-laser impact micro-forming. This innovative method successfully reduced necking, improved residual stress, refined grains, and reduced the rate of surface texture density evolution. In addition to forming force reduction, pulse current–assisted laser shock (EP-LSF) can enhance the corrosion resistance of a material. Liu et al. [25] employed EP-LSF on lra-15 boron nitride anti-oxidation coated AZ31B sheet. The process resulted in a 28.8% increase in forming height, a 6.7% decrease in thinning gradient, and a strain rate–sensitivity coefficient of 0.1452. EP-LSF improved formability by reducing grain size and texture density, while also enhancing corrosion resistance, as evidenced by a decrease in corrosion current density.

Ultrasonic vibration-assisted (UVA) micro forming

Ultrasonic vibration (UV) has been widely used in manufacturing processes [26], and surface treatment methods like ultrasonic shot peening [27]. The UVA micro-forming process has attracted researchers due to its advantages of grain refinement, strain increment, decreased forming force and increased accuracy of micro-holes, improved product hardness, and uniform wall thickness distribution [28]. As presented in Table 1, the amplitude and frequency of vibration are the main process-specific variables, which significantly affect the formability of the component [29]. The schematic diagram of the process is presented in FIG. 4. A positive correlation between amplitude and reduction in forming forces is observed in the micro-forming process. Cheng et al. [30] conducted a study on Al 7075 O-tempered material to investigate the UV effect on incremental forming. They found that the sheet thickness and surface quality were improved in the area where UV was applied. The stress of the material is decreased instantly, and the stress reduction level becomes proportional to the UV amplitude. Cao et al. [31] proved this by investigating the volume and surface effect of 6061 Al alloy. The volume effect results show an instant decrease in material stress that is proportional to the UV amplitude which is 31.67% when the amplitude is 2.96 μm. The surface effect result indicates that UV can refine the grain and increase the proportion of the contact surface's bright area to 32.37% when the amplitude is 2.96 μm. The application of UV in forming processes results in a minimal or insignificant increase in material temperature. During the micro-tension process of T2 copper foil with a thickness of 200 μm, Liu et al. [32] observed an insignificant temperature rise on the formed part, indicating that the thermal effect of UV can be disregarded. Additionally, it was found that the utilization of UV resulted in a decrease of stress by 30% in both micro punching and microstamping [33]. The UV effectiveness may be affected by the geometry of the tool. Zhang et al. [34] studied the effects of tool parameters on forming force, sheet thickness reduction, and spring back. They found that a larger tool radius improves thickness distribution but reduces UV effectiveness. Increasing vibration amplitude, tool diameter, and revolution speed refine the forming process. However, higher feed rates and pitch sizes worsen thickness distribution; in addition larger tool diameters or higher feed rates increases the spring-back effect.

Materials Joining and Manufacturing Processes: MJMP 2025   Materials Research Forum LLC
Materials Research Proceedings 55 (2025) 85-94   https://doi.org/10.21741/9781644903612-14

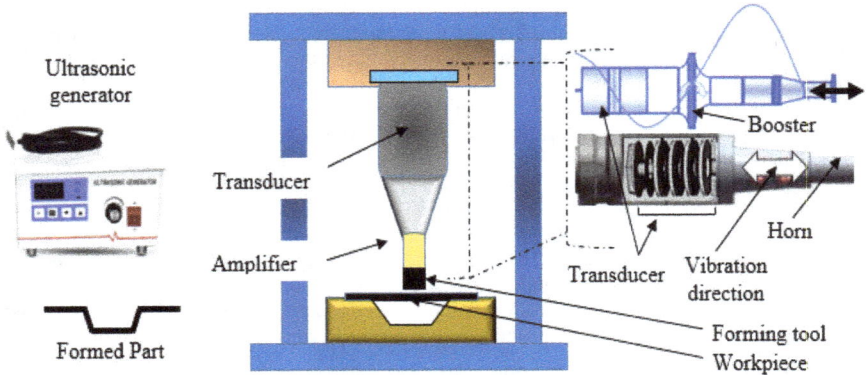

FIG. 4. UVA forming schematic diagram (a) Controller, and (b) Forming system

A rapid drop in stress in the Inconel 718 sheet was observed upon UV superimposition, consistently applied within a strain range of 1.5% to 4.5% to maintain stability [35]. If the experimental amplitude exceeds a critical threshold, the UV system may become overloaded and shut down automatically. Additionally, the force reduction and material formability exhibit a direct proportionality to the UV amplitude [36].

Table 1: Influencing parameters in UV-assisted micro-forming

Material	Size (mm)	Amplitude (μm)	f (kHz)	Findings	Ref.
Inconel 718	0.12	0,1.66,3.18,4.58	-	Flow stress (Yf) =12.9 MPa under (1.66 μm to 4.58 μm) and E = 20.7% to 25.5% under (3.18 μm)	[35]
T2 pure copper	Ø5	7, 9, 11, 13, 15	15Hz	The rate of growth in the height of the micro-cylinder is directly ∝ to the ultrasonic amplitude.	[36]
Pure Copper	Ø2	1, 2, 3.21	60	With higher amplitude (A), the stress reduction becomes more significant.	[37]
Copper	0.2	0.5,1,1.5,2,2.5, 3, 3.5, 4, 5, 6	60	UV effect influences the surface roughness more significantly than acoustic softening.	[38]
Al alloy GB 5052	0.05-0.1	0.48-15	21.1	UVA tests decrease yield stress and UTs, with the effect reversing when deactivated	[39]
Ti3Al	Ø 15	0, 1.31, 3.93, 6.55	0, 20	HV reduction was 6, 3, and 1% under 1.31, 3.93 and 6.55 μm, σ_{max} reduction is 206.13 MPa (6.55 μm).	[40]
TA1	20	4.54,6.02,6.98, 8.46, 10.3	19.846	The optimized parameters to form large-sized micro-grooves with high quality are leading angle = 60°, extruded angle = 40°, and A = 10.30 μm.	[41]
AA5052-H32	1	0-15	20	An increase in the rotational tool speed and vibrating amplitude decreases forming forces.	[42]
Zr-Based alloys	Ø2	0-50	20	UV exposure increases rejuvenation, while ultrasonic intermittent time does not affect crystallization.	[43]
SUS304	0.1, 0.2	3.9, 7.8, 10.1	19.891	The friction coefficient decreases with the increase of ultrasonic vibration amplitude.	[44]

Although UV-assisted microforming does not cause overheating or a significant temperature increase in the specimen, prolonged application of ultrasonic vibrations can lead to excessive

Materials Joining and Manufacturing Processes: MJMP 2025　　　　Materials Research Forum LLC
Materials Research Proceedings 55 (2025) 85-94　　　　https://doi.org/10.21741/9781644903612-14

heating and a notable rise in the specimen's temperature [36]. To mitigate the ultrasonic surface effect caused by re-lubrication, it is recommended to avoid using lubricant [37]. The research underscores the potential of UV-assisted forming in achieving high-quality micro-scale products. By leveraging high-frequency vibrations, this technique reduces forming forces, improves material flow, and enhances surface quality. While challenges remain, it holds significant promise for advancing micro-manufacturing technologies [36], [37], [39].

Future research on aforementioned emerging micro-forming methods should focus on standardizing testing protocols to ensure consistency and comparability across studies. Multi-scale modeling approaches are needed to link microstructural mechanisms with macroscopic behavior, incorporating factors like thermal effects and material flow. Expanding the study to include diverse materials and assessing long-term product performance under various conditions will enhance understanding. Additionally, integrating sustainability metrics will help evaluate the environmental and energy efficiency of the process, ensuring its broader adoption in precision manufacturing.

Conclusion

This review analyzed recent developments in micro-forming processes for fabricating intricate shapes, identifying key parameters that influence method performance, achievements, and research gaps. EAF enhances plasticity and controls spring-back, with successful applications in various materials, while UVA forming improves material properties and uniform wall thickness. Laser shock forming induces favorable microstructural changes, enhances material strength, and enables precise deformation control. The review highlights the impact of process parameters on micro-forming efficiency and discusses gaps in understanding material behavior at the microscale. Future research should focus on advanced material characterization, multi-scale modeling, process optimization, and exploring new materials and applications. Addressing these gaps will drive advancements in precision manufacturing through micro-forming.

Funding

The author declares that no funding was received for this study.

Conflict of Interest

The author declares no conflict of interest.

References

[1] A. Ivanov, K. Cheng, Non-traditional and hybrid processes for micro and nano manufacturing, Int. J. Adv. Manuf. Technol. 105 (2019) 4481–4482. https://doi.org/10.1007/S00170-019-04711-0

[2] G. Patel, K. Ganesh M, O. Kulkarni, Experimental and numerical investigations on forming limit curves in micro forming, Adv. Mater. Process. Technol. 8 (2022) 1–12. https://doi.org/10.1080/2374068X.2020.1793268

[3] P. Zhang, M. P. Pereira, B. F. Rolfe, D. E. Wilkosz, P. Hodgson, M. Weiss, Investigation of material failure in micro-stamping of metallic bipolar plates, J. Manuf. Process. 73 (2022) 54–66. https://doi.org/10.1016/J.JMAPRO.2021.10.044

[4] H. Zhang, X. Wang, Y. Ma, X. Gu, J. Lu, K. Wang, H. Liu, Formability and mechanism of pulsed current pretreatment–assisted laser impact microforming, Int. J. Adv. Manuf. Technol. 114 (2021) 1011–1029. https://doi.org/10.1007/s00170-021-06964-0

[5] Z. Xu, T. Jiang, J. Huang, L. Peng, X. Lai, M. W. Fu, Electroplasticity in electrically-assisted forming: Process phenomena, performances and modelling, Int. J. Mach. Tools Manuf. 175 (2022) 103871. https://doi.org/10.1016/j.ijmachtools.2022.103871

[6] L. Xu, Y. Lei, H. Zhang, Z. Zhang, Y. Sheng, G. Han, Research on the Micro-Extrusion Process of Copper T2 with Different Ultrasonic Vibration Modes. 9 (2019) 1209. https://doi.org/10.3390/MET9111209

[7] D. Meng, J. Ma, X. Zhao, Y. Guo, C. Zhu, M. Yu, Mechanical behavior and material property of low-carbon steel undergoing low-frequency vibration-assisted upsetting, J. Mater. Res. Technol. 16 (2022) 1846–1855. https://doi.org/10.1016/J.JMRT.2021.12.113

[8] C. Wang, L. Cheng, Y. Liu, C. Zhu, Ultrasonic flexible bulging process of spherical caps array as surface texturing using aluminum alloy 5052 ultra-thin sheet, J. Mater. Process. Technol. 284, (2020) 116725. https://doi.org/10.1016/J.JMATPROTEC.2020.116725

[9] Q. Gong, X. Wang, D. Zhang, X. Hou, T. Zhang, H. Liu, Flow-through and forming mechanism of laser shock micro-coining, J. Mater. Process. Technol. 307 (2022) 117678. https://doi.org/10.1016/J.JMATPROTEC.2022.117678

[10] K. Wang, H. Liu, Y. Ma, J. Lu, X. Wang, J. Lu, X. Gu, H. Zhang, Laser shock micro-bulk forming: Numerical simulation and experimental research, J. Manuf. Process. 64 (2021) 1273–1286. https://doi.org/10.1016/J.JMAPRO.2021.02.049

[11] R. Fabbro, J. Fournier, P. Ballard, D. Devaux, J. Virmont, Physical study of laser-produced plasma in confined geometry, J. Appl. Phys. 68 (1990) 775–784. https://doi.org/10.1063/1.346783

[12] X. Wang, X. Hou, D. Zhang, Q. Gong, Y. Ma, Z. Shen, H. Liu, Research on warm laser shock sheet micro-forging, J. Manuf. Process. 84 (2022) 1162–1183, Dec. 2022. https://doi.org/10.1016/J.JMAPRO.2022.10.076

[13] X. Hou, Y. Ma, X. Wang, W. Shen, M. Cui, H. Liu, Size effect of laser shock sheet bulk microforging, Int. J. Adv. Manuf. Technol. 128 (2023) 2719–2737. https://doi.org/10.1007/S00170-023-12094-6

[14] J. M. López, D. Munoz-Martin, J.J. Moreno-Labella, M. Panizo-Laiz, G. Gomez-Rosas, C. Molpeceres, M. Morales, Picosecond Laser Shock Micro-Forming of Stainless Steel: Influence of High-Repetition Pulses on Thermal Effects, Materials (Basel). 15 (2022). https://doi.org/10.3390/ma15124226

[15] W. A. Salandro, C. J. Bunget, L. Mears, A thermal-based approach for determining electroplastic characteristics. 226 (2012) 775–788. https://doi.org/10.1177/0954405411424696

[16] X. Bing-hui Xing, H. Tao, S. Ke-xing, X. Liu-jie, Y. Si-liang, X. Nan, C. Fu-xiao, Effect of electric current on mechanical properties and microstructure of Ti/Al laminated composite during electrically assisted tension, Vacuum. 210 (2023) 111805. https://doi.org/10.1016/j.vacuum.2023.111805

[17] X. Zhenhai, X. Shaoxi, W. Chunju, W. Xinwei, X. Jie, S. Debin, G. Bin, Electrically assisted micro-rolling process of surface texture on T2 copper sheets, Int J Adv Manuf Technol. 118 (2022) 2453–2466. https://doi.org/10.1007/s00170-021-08094-z1

[18] J. Bao, S. Lv, B. Wang, D. Shan, B. Guo, J. Xu, The Effects of Geometry Size and Initial Microstructure on Deformation Behavior of Electrically-Assisted Micro-Compression in Ti-6Al-4V Alloy, Mater. 15 (2022) 1656. https://doi.org/10.3390/MA15051656

[19] S. Xu, X. Xiao, H. Zhang, Z. Cui, Electroplastic Effects on the Mechanical Responses and Deformation Mechanisms of AZ31 Mg Foils, Mater. 15 (2022) https://doi.org/10.3390/MA15041339

[20] B. Jiang, W. Yang, Z. Zhang, X. Li, X. Ren, Y. Wang, Numerical Simulation and Experiment of Electrically-Assisted Incremental Forming of Thin TC4 Titanium Alloy Sheet, Mater. 10 (2020) 1335. doi: https://doi.org/10.3390/MA13061335

[21] Y. Zhou, C. Wu, Z. Qu, B. Lin, Ti6554 titanium alloy electrically assisted compression: modelling and simulation based on dislocation density theory, IOP Conf. Ser. Mater. Sci. Eng. 1270 (2022) 012065. https://doi.org/10.1088/1757-899X/1270/1/012065

[22] Q. Liu, W. Fan, Z. Ren, T. Wang, Q. Huang, Effect of Pulse Current Treatment on Microstructure and Properties of 304 Stainless Steel Ultra-thin Strip after Rolling, Chin. J. Mech. 37 (2024) 149. https://doi.org/10.1186/s10033-024-01119-0

[23] O. Erdene-Ochir, J. Liu, D. M. Chun, Effect of pulsed electric current on electrically assisted indentation for surface texturing, Int. J. Adv. Manuf. Technol. 111 (2020) 283–293. https://doi.org/10.1007/S00170-020-06102-2

[24] Y. Sun, X. Wang, Y. Ma, H. Liu, Investigation on improved micro-formability and forming mechanism under high strain rate of H62 brass in pulsed current-assisted laser impact micro-forming, Mater. Sci. Eng. A. 830 (2022) 142301. https://doi.org/10.1016/j.msea.2021.142301

[25] H. Liu, Y. Sun, Y. Ma, Y. He, X. Wang, Improvement of formability and corrosion resistance of AZ31 magnesium alloy by pulsed current–assisted laser shock forming, Int. J. Adv. Manuf. Technol. 120 (2022) 6531–6545. https://doi.org/10.1007/S00170-022-09211-2

[26] A. I. Gorunov, O. A. Nyukhlaev, A. K. Gilmutdinov, Investigation of microstructure and properties of low-carbon steel during ultrasonic-assisted laser welding and cladding, Int. J. Adv. Manuf. Technol. 99 (2018) 2467–2479. https://doi.org/10.1007/S00170-018-2620-7

[27] M. G. Ganta and M. Kurek, "Influence of post-processing methods on the fatigue performance of materials produced by selective laser melting (SLM)," Int. J. Adv. Manuf. Technol. 136 (2025), 2003–2034. https://doi.org/10.1007/S00170-024-14920-X

[28] S. Kumar, D. Kumar, I. Singh, D. Rath, An insight into ultrasonic vibration assisted conventional manufacturing processes: A comprehensive review, Adv. Mech. Eng. 14 (2022). https://doi.org/10.1177/16878132221107812

[29] A. Gohil, B. Modi, K. Patel, Effect of amplitude of vibration in ultrasonic vibration-assisted single point incremental forming, Mater. Manuf. Process. 37 (200) 1837–1849. https://doi.org/10.1080/10426914.2022.2065008

[30] R. Cheng, N. Wiley, M. Short, X. Liu, and A. Taub, Applying ultrasonic vibration during single-point and two-point incremental sheet forming, Procedia Manuf. 34 (2019)186–192. https://doi.org/10.1016/J.PROMFG.2019.06.137

[31] M. Y. Cao, H. Hu, X. D. Jia, S. J. Tian, C. C. Zhao, X. B. Han, Mechanism of ultrasonic vibration assisted upsetting of 6061 aluminum alloy, J. Manuf. Process. 59 (2020) 690–697. https://doi.org/10.1016/J.JMAPRO.2020.09.070

[32] Y. Liu, C. Wang, R. Bi, Acoustic residual softening and microstructure evolution of T2 copper foil in ultrasonic vibration assisted micro-tension, Mater. Sci. Eng. A. 841 (2022) 143044. https://doi.org/10.1016/j.msea.2022.143044

[33] G. Kiswanto1, H. T. Tjong, S. T. Dwiyati, S. Supriadi, W. Z. Abdurrohman, E. J. P. Mendrofa, R A Adityawarman, Development of progressive tool system for ultrasonic vibration assisted microforming, Mater. Res. Proc. 25 (2023) 297–304. https://doi.org/10.21741/9781644902417-37

[34] L. Zhang, C. Wu, H. Sedaghat, Ultrasonic vibration–assisted incremental sheet metal forming, Int. J. Adv. Manuf. Technol. 114 (2021) 3311–3323. https://doi.org/10.1007/S00170-021-07068-5

[35] B. Meng, B. N. Cao, M. Wan, C. J. Wang, D. B. Shan, Constitutive behavior and microstructural evolution in ultrasonic vibration assisted deformation of ultrathin superalloy sheet, Int. J. Mech. Sci. 157 (2019) 609–618. https://doi.org/10.1016/J.IJMECSCI.2019.05.009

[36] J. Zhai, Y. Guan, W. Wang, L. Zhu, Z. Xie, J. Lin, Studies on ultrasonic vibration-assisted coining of micro-cylinder, Int. J. Adv. Manuf. Technol. 100 (2019) 2031–2044. https://doi.org/10.1007/S00170-018-2806-Z

[37] J. Hu, T. Shimizu, T. Yoshino, T. Shiratori, M. Yang, Evolution of acoustic softening effect on ultrasonic-assisted micro/meso-compression behavior and microstructure, Ultrasonics, 107 (2020) 106107. https://doi.org/10.1016/J.ULTRAS.2020.106107

[38] Z. Yin, M. Yang, Investigation on Deformation Behavior in the Surface of Metal Foil with Ultrasonic Vibration-Assisted Micro-Forging, Mater. 15 (2022) 1907 https://doi.org/10.3390/MA15051907

[39] C. Wang, W. Zhang, L. Cheng, C. Zhu, X. Wang, H. Han, H. He, R. Hua, Investigation on Microsheet Metal Deformation Behaviors in Ultrasonic-Vibration-Assisted Uniaxial Tension with Aluminum Alloy 5052, Mater. 13 (2020) 637. https://doi.org/10.3390/MA13030637

[40] Z. Fu, G. Gao, Y. Wang, D. Wang, D. Xiang, B. Zhao, Investigation of acoustic softening and microstructure evolution characteristics of Ti3Al intermetallics undergoing ultrasonic vibration-assisted tension, Mater. Des. 222 (2022) 111015. https://doi.org/10.1016/J.MATDES.2022.111015

[41] A. N. N. Kakahy, W. F. A. Alshamary, A. A. Kakei, H. D. Zhang, X. Y. Wang, Research on ultrasonic vibration assisted ploughing-extrusion process in forming μ-scale micro grooves, IOP Conf. Ser. Mater. Sci. Eng. 1270 (2022) 012104. https://doi.org/10.1088/1757-899X/1270/1/012104

[42] T. K. Le, D. T. Tran, N. T. B. Bui, T. H. Nguyen, Effect of Ultrasonic Vibration on Forming Force in the Single-Point Incremental Forming Process, Shock Vib. (2023). https://doi.org/10.1155/2023/1565927

[43] Y. Lingyun, X. Shenpeng, L. Yan, Effect of Intermittent Ultrasonic Vibration-Assisted Compression on the Mechanical Properties of Zr-Based Amorphous Alloys, Front. Mater. 8 (2021) 801991. https://doi.org/10.3389/FMATS.2021.801991/BIBTEX

[44] C. Zha and S. Zha, "Theoretical and Experimental Study on the Influence of Ultrasonic Vibration on Contact Friction, Manufacturing Technology, 22 (2022) 367–376. https://doi.org/10.21062/MFT.2022.038

Materials Joining and Manufacturing Processes: MJMP 2025
Materials Research Proceedings 55 (2025) 95-101

Materials Research Forum LLC
https://doi.org/10.21741/9781644903612-15

Influence of tool rotation speed on mechanical and microstructural characteristics of friction stir weld 7075 aluminum alloy reinforced with graphene nano platelets

Rahul Biradar[1,a] * and Sachinkumar Patil[1,b]

[1]School of Mechanical Engineering, REVA University Bengaluru, Karnataka, India

[a]rahulbiradar100@gmail.com, [b]sachindongapur@gmail.com

Keywords: AA7075, Friction Stir Welding, Graphene Nanoplatelets, Mechanical Properties, Microstructure

Abstract. The growing demand for lightweight, durable materials within the aerospace industry has motivated researchers to develop composites exhibiting exceptional mechanical properties, thereby enabling a more economical and environment-friendly transformation. Carbonaceous reinforcement is increasingly prioritized due to its exceptional properties in modifying and enhancing the mechanical, microstructure, and tribological properties. In the present research, the nanocomposite of AA7075 aluminum alloy reinforced with graphene nanoplatelets (GNPs) has been developed through friction stir welding (FSW) technique. Obtained the significant process parameter of tool rotation speed 800 rpm, traverse speed 20 mm/min, and tilt angle 2° respectively. Microstructure in nugget zone exhibits fine and uniform grains. It results in significant enhancement of ultimate tensile strength (UTS) 821 MPa and yield strength (YS) 389 MPa. Further, higher hardness of 141 HV is reported in the nugget zone. Moreover, the joint is fractured at nugget zone and it represents the weakest part of the weld joint.

Introduction

The integration of two different materials to form a combined system is referred as composite materials. The primary materials are considered as the matrix, while supplimentary materials are designated as reinforcement. A major benefit of composite materials as compared to alloys their ability to incorporate anisotropic characteristics. It results in diverse physical and mechanical characteristics of aluminum alloys such as lightweight and good formability [1, 2]. Aluminum and its alloys are broadly used from household utensils to aerospace structures. Specially in aircraft structure upper and lower wing parts are used [3]. The aluminum alloys are employed in structures that are predominantly joined by using rivets, which contribute to an increase of overall weight of the aircraft [4]. Therefore, friction stir welding (FSW) process operates below the melting point of temperature is to be welded. However, few defects are identified such as crack propagation and excess slag in conventional joints. Therefore solid-state welding technique named FSW is developed by The Welding Institute (TWI) [5, 6]. In FSW process, non-consumable rotating tool is equipped with pin and its shoulder employed to facilitate joint formation by the abutting edge of the plates, which results in transverse along the weld joint [7, 8]. The FSW emerges as an ingenious solid-state welding technique, it offers numerous advantages when employed to join high-resistance 7xxx series aluminum alloys as compared to traditional fusion welding technique [9].

Moreover, AA7075 is renowned for its superior properties it has been extensively employed in aircraft components and highly demanded various applications such as marine, automotive, and structural applications [10, 11]. Several researchers focused on the influence of various parameters like pin profile, tool rotational speed (TRS), and tool traverse speed (TTS). Therefore, few relevent studies are discussed here. Fruitful investigations by Khodir et al. [12] reported significant

Content from this work may be used under the terms of the Creative Commons Attribution 3.0 license. Any further distribution of this work must maintain attribution to the author(s) and the title of the work, journal citation and DOI. Published under license by Materials Research Forum LLC.

mechanical properties are obtained by AA2024 and AA7075 aluminum alloy. Guo et al. [13] reported FSW of AA6061 and AA7075 dissimilar materials. The material mixing is more effective in advancing side. The UTS of FSW of AA6061 and AA7075 joints is found to be increased as compared to base material. Sivaraman et al. [14] studied FSW of AA2014 and AA7075 alloy with a rotational speed of 1000 rpm. It results in good weld surface and sound weld joints. The highest UTS is achieved at 160 MPa. Tamilselvan et al. [15] reported FSW of AA6061 and AA7075 alloy, the joint is produced by better wettability and higher hardness at spindle speed of 710 rpm and traverse speed of 14 mm/sec. Martin et al. [16] reported FSW of AA7075-T651 alloy. It results incomplete recrystallization occurs in the NZ. Further, the YS and UTS were obtained by 55 % and 74 % respectively. Chenghang et al. [17] investigated FSW of AA2024/AA7075 alloy, the material is mixed properly due to tool shoulder design, tool pin, and transition between the shoulder and pin. The maximum tensile strength is recorded at 446 MPa for rolling and transverse direction. Navdeep et al. [18] investigated FSW of AlSi10Mg and AA7075 alloys fabricated with higher heat input. It results maximum hardness is achieved in the nugget zone. Its owing to MgZn$_2$ and β-Mg$_2$Si secondary phase particles are observed. Moreover, fine grains and higher dislocation density were noticed in the weld region. It achieved maximum joint efficiency of 77 %. Ahmed et al. [19] investigated FSW of AA7075 and AA5083 dissimilar alloys, significant grain refinement takes place in the nugget zone with average grain size (AGS) of 6 μm than base material. The hardness profile of dissimilar joints produced a smooth transition. The fracture surface reveals the mixed mode of failure. Edip et al. [20] studied FSW of AA7075/AA5182 alloy, significant mechanical properties were obtained by conical shape tool pin. The tensile and fatigue strength are obtained by 265 MPa and 159 MPa respectively. The higher hardness is determined as 87 HV in NZ at rotational speed of 1325 rpm. The current investigation systematically reported the influence of different TRS such as 700, 800, 900, and 1000 rpm and constant traverse speed of 20 mm/min on microstructure and mechanical characteristics of friction stir welded (FSWed) graphene nanoplatelets reinforced with AA7075.

Experimental procedure

The AA7075/GNPs composite of 6 mm thickness plate were welded by FSW set up (Three-axis, ETA Private Limited Bangalore) under various TRS of 700, 800, 900, and 1000 rpm and traverse speed of 20 mm/min. The chemical composition of base material is evaluated in weight percentage and represented in Table 1. The mechanical characteristics of base material also evaluated and represented in Table 2. A square type of flat-faced H13 steel is used. The tool is rotated in clockwise direction. The microstructure of base material and FSWed joints is characterized by optical microscopy (OM). The specimens were cut from the transverse weld directions. The microstructure samples were polished with different grades of emery sheets such as 100, 200, 300, and 400 silicon carbide (SiC) papers. After that alumina disc polish performed with considering micro size alumina powder. In continuation to disc polishing, the diamond polished were performed for more surface finishing purpose. Further, FSWed joints are shown in Fig. 1 (a), also microstructure and microhardness samples are shown in Fig. 1 (b).

Table 1: Chemical composition of AA7075.

Element	Zn	Mg	Mn	Cu	Si	Fe	Ti	Cr	Al
Weight %	5.6	2.5	0.3	1.5	0.4	0.5	0.2	0.25	91.5

Table 2: Mechanical properties of AA7075 aluminum alloy.

Material	UTS (MPa)	YS (MPa)	Elongation (%)
AA7075	175	125	18.33

Materials Joining and Manufacturing Processes: MJMP 2025   Materials Research Forum LLC
Materials Research Proceedings 55 (2025) 95-101   https://doi.org/10.21741/9781644903612-15

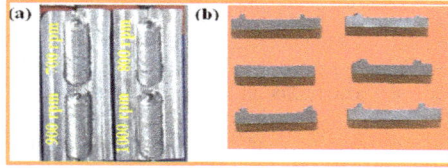

Fig. 1: (a) FSWed joints and (b) Microstructure and microhardness samples.

The mechanical characteristics of tensile and microhardness tests were performed. The microhardness test (Model: Wilson VH1102) is conducted under a load of 1 Kgf and maintained at a time of 10 s. Tensile samples are produced according to ASTM E 08 sub-size dimensions. The tensile test (Model: Biss Instron) is carried out at room temperature. Tensile samples before fracture and after fracture as exhibits in Fig. 2 (a) and (b) respectively.

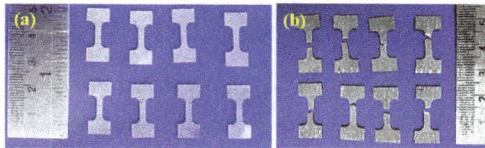

Fig. 2: (a) Tensile samples before fracture (b) Tensile samples after fracture.

Results and Discussions

Macrostructure observation

Fig. 3 presents the macroscopic observation of FSWed cross-section samples. In fusion welding aluminum alloys are produced flaws such as cracks, slag inclusion and porosity. These defects affects the quality of weld joint. Therefore, FSW is free from all these defects. From, Fig. 3 its observed that there is no defects occured in FSWed cross section samples. Therefore, it indicates that sound weld joint is formed. These macrographs are reveals the basin shape in nugget region. Its considered as common features in FSW process and similar results are reported by several researchers [21].

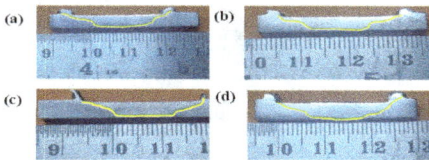

Fig. 3: Macroscopic observation of FSWed samples at constant traverse speed of 20 mm/min and various TRS of (a) 700, (b) 800, (c) 900, and (d) 1000 rpm.

Microstructure observation

Fig. 4 exhibits base material it consists of large and elongated grain structure. In FSW process, the nugget zone (NZ) is characterized by the presence of fine and equiaxed grains as compared to various zones. It's due to frictional heat and subsequent plastic deformation results from the movement of material by the rotating tool pin. The FSW process is recognized is having high strain rates under severe plastic deformation. This process is called as dynamic recrystallization (DRX) [22, 23]. During DRX the occurrence of nucleation serves as barrier to grain boundary movement.

In general, the significant efficiency of dynamic recovery combined with the reduced mobility of grain boundaries facilitates continuous DRX in the NZ as shown in Fig. 5 (a to d), the presence of greater density is responsible for the decreased mobility of grain boundaries in NZ. The average grain size (AGS) is found 6 to 8 μm. In thermo mechanical affected zone (TMAZ) the grains are elongated. In this region, the material is plastically deformed. The AGS is observed as 12 to 14 μm. In heat affected zone (HAZ), grains are not affected by the heat and frictional force. It's because of material undergoes a thermal state which increases the grain size. The AGS is observed as 20 to 22 μm in HAZ.

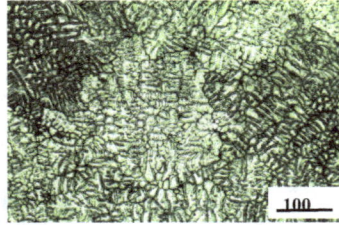

Fig. 4: Base material AA7075/GNPs composites.

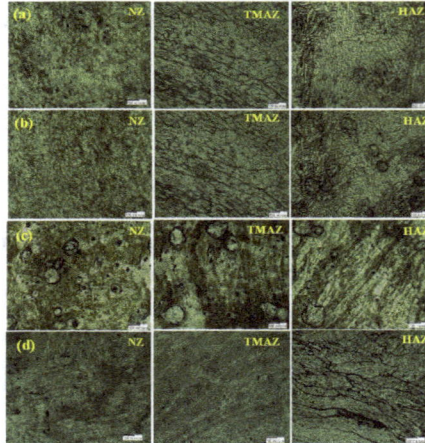

Fig. 5: Optical photomicrograph of AA7075/GNPs composites at different TRS and constant traverse speed of 20 mm/min (a) 700, (b) 800, (c) 900, and (d) 1000 rpm.

Tensile and microhardness characterization

Fig. 6 (a) illustrates the stress-strain diagram of various TRS such as 700, 800, 900, and 1000 rpm and traverse speed of 20 mm/min. The UTS is decreases due to higher rotational speed. It's because of material is not mixed properly due to lower heat input. From, Fig. 6 (a) it's clear that 800 rpm weld joints exhibit higher UTS : 821 MPa, YS: 389 MPa and % of elongation: 21% as compared to various TRS. Minimum UTS and YS is obtained by 1000 rpm. Moreover, in HAZ the material remains unaffected by mechanical stirring process. Therefore, failure occurs in the weld joint due

Materials Joining and Manufacturing Processes: MJMP 2025   Materials Research Forum LLC
Materials Research Proceedings 55 (2025) 95-101   https://doi.org/10.21741/9781644903612-15

to thermal softening. For 900 rpm obtained the lower UTS: 527 MPa, YS: 314 MPa, and % of elongation: 21%. Its due to improper material mixing as shown in Fig. 5 (c). Further, similar findings are evidenced by Harish et al. [24] reported failure occurs in HAZ due to the presence of defects. It results in lower elongation of the joint.

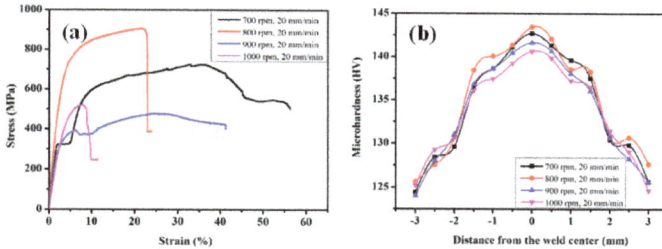

Fig. 6: Different TRS and constant traverse speed of 20 mm/min (a) Stress-strain diagram and (b) microhardness of FSWed joint cross-section

Vicker microhardness test is carried out across the joint cross-section to study the microhardness and noticed variations in different regions of the weldments. The microhardness distribution is exhibited in Fig. 6 (b). In 7xxx series, alloys are welded under conditions for which the nugget zone temperature is approximated to the solution heat treatment. In weld nugget zone the hardness is observed to be 143 HV for 800 rpm. It's due to significant grain refinement occurres in the tool stirring of the weld joint. According to the Hall-Petch effect, the hardness value exhibits an inverse proportional to the grain size. Grain refinement results from intense plastic deformation, which leads to greater hardness in the nugget zone. Moreover, lower hardness is obtained in HAZ, due to the presence of coarser grains. Further, similar results are obtained by Chenghang et al. [17] in nugget zone obtained the higher microhardness than base material.

Conclusions
In the current study, the AA7075/GNPs composites welded by friction stir welding process. The traverse speed maintained constant 20 mm/min with different TRS of 700, 800, 900, and 1000 rpm. An investigation of mechanical and microstructural properties were evaluated, leading to the derivation of the following conclusions.

1. As the TRS gradually increases the hardness is decreases. The lower hardness is recorded in TMAZ and HAZ. The maximum hardness is reached to 143 HV in nugget zone at TRS of 800 rpm.
2. The TRS of 800 rpm results significant UTS of 821 MPa and YS of 389 MPa respectively.
3. Maximum tensile samples are fractured in HAZ due to the presence of coarser precipitates. In HAZ the tool is not stirred significantly. Therefore, failure occurs in ductile mode.
4. The TRS of 800 rpm exhibits fine grains in nugget region as compared to various TRS, with AGS of 6 μm.

Acknowledgements
The authors wish to extend our appreciation to REVA University Bangalore for the financial support and facilities for conducting the research [Ref No: RU: EST: ME: 2022-1].

References

[1] S.T.A. Filho, S. Sheikhi, J.F. Santos, C. Bolfarini, Preliminary study on the microstructure and mechanical properties of dissimilar friction stir welds in aircraft aluminium alloys 2024-T351 and 6056-T4, J. Mat. Pro. Tech. 206 (2008) 132-142.
https://doi.org/10.1016/j.jmatprotec.2007.12.008

[2] N.A.A. Sathari, L.H. Shah, A.R. Razali, Investigation of single-pass/double-pass techniques on friction stir welding of aluminium, J. Mech. Eng. Sci. 7 (2014) 1053-1061.
http://dx.doi.org/10.15282/jmes.7.2014.4.0102

[3] L. H. Shah, Z. Akhtar, M. Ishak, Investigation of aluminum-stainless steel dissimilar weld quality using different filler metals, Int. J. Auto. Mech. Eng. 8 (2013) 1121-1131.
http://dx.doi.org/10.15282/ijame.8.2013.3.0091

[4] P. Sachinkumar, S. Narendranath, D. Chakradhar, Microstructure, hardness and tensile properties of friction stir welded aluminum matrix composite reinforced with SiC and Fly Ash, Silicon. 11 (2019) 2557–2565. https://doi.org/10.1007/s12633-018-0044-5

[5] Z.W Zhang, Z.Y Liu, B.L Xiao, D.R Ni, Z.Y Ma, High efficiency dispersal and strengthening of graphene reinforced aluminum alloy composites fabricated by powder metallurgy combined with friction stir processing, Carbon. 135 (2018) 215-223.
https://doi.org/10.1016/j.carbon.2018.04.029

[6] R. Biradar, S. Patil, A systematic review on microhardness, tensile, wear, and microstructural properties of aluminum matrix composite joints obtained by friction stir welding: past, present and its future, Tran. Ind. Inst. Met. 77 (2024) 1923-1937.
https://doi.org/10.1007/s12666-024-03303-1

[7] S.M. Bayazid, H. Farhangi, A. Ghahramani, Investigation of friction stir welding parameters of 6063-7075 aluminum alloys by Taguchi method, Pro. Mat. Sci. 11 (2015) 6-11.
https://doi.org/10.1016/j.mspro.2015.11.007

[8] J.A. Hamed, Effect of welding heat input and post-weld aging time on microstructure and mechanical properties in dissimilar friction stir welded AA7075−AA5086, Tran. Non. Met. Soc. China. 27 (2017) 1707−1715. https://doi.org/10.1016/S1003-6326(17)60193-6

[9] M. Rashad, F. Pan, A. Tang, M. Asif, Effect of GNPs addition on mechanical properties of pure aluminum using a semi-powder method, Prog. Natu. Sci: Mat. Int. 24 (2014) 101-108.
https://doi.org/10.1016/j.pnsc.2014.03.012

[10]F. Zhiyuan, L. Jiao, M. Jincai, S. Yongjin, Z. Xiaoyuan, M. Yu, Z. Zilong, EBSD characterization of 7075 aluminum alloy and its corrosion behaviors in SRB marine environment, J. Mar. Sci. Eng. 10 (2022) 1-10. https://doi.org/10.3390/jmse10060740

[11]P. Sachinkumar, S. Narendranath, D. Chakradhar, Characterization and evaluation of joint properties of FSWed AA6061/SiC/FA hybrid AMCs using different tool pin profiles, Tran. Ind. Inst. Met. 73 (2020) 2269 – 2279. https://doi.org/10.1007/s12666-020-02035-2

[12]S.A. Khodir, T. Shibayanagi, Friction stir welding of dissimilar AA2024 and AA7075 aluminum alloys, Mat. Sci. Engg B. 148 (2008) 82-87.
https://doi.org/10.1016/j.mseb.2007.09.024

[13]J.F. Guo, H.C. Chen, C.N. Sun, G. Bi, Z. Sun, J. Wei, Friction stir welding of dissimilar materials between AA6061 and AA7075 Al alloys effects of process parameters, Mat. Des. 56 (2014) 185-192. http://dx.doi.org/10.1016/j.matdes.2013.10.082

[14] P. Sivaraman, T. Nithyanandhan, M. Karthick, S.M. Kirivasan, S. Rajarajan, M. Sivanesa Sundar, Analysis of tensile strength of AA 2014 and AA 7075 dissimilar metals using friction stir welding, Mat. Today. Proc. 37 (2021) 187-192. https://doi.org/10.1016/j.matpr.2020.04.895

[15] T. Amuthan, N. Nagaprasad, R. Krishnaraj, V. Narasimharaj, B. Stalin, V. Vignesh, Experimental study of mechanical properties of AA6061 and AA7075 alloy joints using friction stir welding, Mat. Today. Proc. 47 (2021) 4330-4335. https://doi.org/10.1016/j.matpr.2021.04.628

[16] M. Reimann, J. Goebel, J F. Santos, Microstructure and mechanical properties of keyhole repair welds in AA 7075-T651 using refill friction stir spot welding, Mat. Des. 132 (2017) 283-294. http://dx.doi.org/10.1016/j.matdes.2017.07.013

[17] C. Zhang, G. Huang, Y. Cao, Y. Zhu, W. Lia, X. Wang, Q. Liu, Microstructure and mechanical properties of dissimilar friction stir welded AA2024-7075 joints: Influence of joining material direction, Mat. Sci. Eng A. 766 (2019) 138368. https://doi.org/10.1016/j.msea.2019.138368

[18] N. Minhas, V. Sharma, S. Manda, A. Thakur, Insights into the microstructure evolution and mechanical behavior of dissimilar friction stir welded joints of additively manufactured AlSi10Mg and conventional 7075-T651 aluminum alloys, Mat. Sci. Eng A. 881 (2023) 145407. https://doi.org/10.1016/j.msea.2023.145407

[19] A. Hamdollahzadeh, M. Bahrami, M.F. Nikoo, A. Yusefi, M.K.B. Givi, N. Parvin, Microstructure evolutions and mechanical properties of nano-SiC-fortified AA7075 friction stir weldment: The role of second pass processing, J. Manu. Pro. 20 (2015) 367-373. http://dx.doi.org/10.1016/j.jmapro.2015.06.017

[20] E. Cetkin, Y.H Çelik, S. Temiz, Microstructure and mechanical properties of AA7075/AA5182 jointed by FSW, J. Mat. Proc. Tech. 268 (2019) 107-116. https://doi.org/10.1016/j.jmatprotec.2019.01.005

[21] M. Simoncini, A. Forcellese, Effect of the welding parameters and tool configuration on micro and macro mechanical properties of similar and dissimilar FSWed joints in AA5754 and AZ31 thin sheets, Mat. Des. 41 (2012) 50-60. http://dx.doi.org/10.1016/j.matdes.2012.04.057

[22] S.S. Mirjavadi, M. Alipour, S. Emamian, S. Kord, A.M.S. Hamouda, G. Praveennath, R. Keshavamurthy, Influence of TiO2 nanoparticles incorporation to friction stir welded 5083 aluminum alloy on the microstructure, mechanical properties, and wear resistance, J. Allo. Comp. 712 (2017) 795-803. http://dx.doi.org/10.1016/j.jallcom.2017.04.114

[23] B. Singh, K. Saxena, P. Singhal, Role of various tool pin profiles in friction stir welding of AA2024 alloys, J. Mat. Eng. Perf. 30 (2021) 8606–8615. https://doi.org/10.1007/s11665-021-06017-3

[24] H. Suthar, A. Bhattacharya, S. K. Paul, Local deformation response and failure behavior of AA6061-AA6061 and AA6061-AA7075 friction stir welds, CIRP. J. Manu. Sci. Tech. 30 (2020) 12-24. https://doi.org/10.1016/j.cirpj.2020.03.006

Materials Joining and Manufacturing Processes: MJMP 2025   Materials Research Forum LLC
Materials Research Proceedings 55 (2025) 102-109   https://doi.org/10.21741/9781644903612-16

Investigation on mechanical and microstructural properties of friction stir weld AA8xxx series alloy: A review

Maheshwarayya K.C.[1,a], Sachinkumar Patil[1,b], Mahesh L.[1,c]

[1]School of Mechanical Engineering, REVA University Bengaluru, Karnataka, India – 560064

[a]maheshwarayyakc@gmail.com, [b]sachindongapur@gmail.com
[c]maheshl@reva.edu.in

Keywords: AA8009, Friction Stir Welding, Hardness, Tensile Strength, Microstructure

Abstract. The utilization of aluminium alloy in aerospace, automotives, and marine industrial applications is ever increasing. As aluminium is lighter in weight, corrosive resistance and less costly compared to other alloys. A green welding process is named friction stir welding is extremely used for joining similar and dissimilar materials. The process uses non consumable tool to produce frictional heat without melting base material. In FSW process parameters such as tool shoulder diameter, rotational speed, welding speed and tilt angle do major role in obtaining better quality joints. In this review paper FSW joint characteristics of AA 8xxx alloys have been studied and focused on mechanical and microstructural properties. In addition, this study reports the formation fine grains in weld joints. FSW joint exhibits better strength without heat affected zone. Formation of fine grains is noticed in FSW microstructure attributed to stirring action of FSW tool.

Introduction:

Friction stir welding is a solid state of joining process, in which material is being welded does not melt or recast. It is advanced type of metal joining process and more potential for joining aluminium alloy FSW joints significantly reduces the defect like porosity, crack, and distortions occurred by the fusion welded joints. As the aluminium is light weight, corrosive resistance and less costly as compared to other alloys as result aluminium is widely utilized in aerospace industry, automotive industry, high speed ships, railway vehicles, and military applications [1]. FSW have better advantages in mechanical treatment methods for the surface, this makes welding process is versatile and unique welding method [2]. Aluminium alloy 8xxx is thermally stable possess low density, high strength at room temperature. A series which has the most comprehensive properties such as high ductility, fracture, toughness also AA8xxx is high heat resistance aluminium alloy, has high temperature strength than any other conventional aluminium alloys. The effective use temperature range is extended more than 300° C therefore, 8xxx aluminium alloy is capable to replace the traditional heat resistance alloy at 300 – 400° C this reasons the dominant titanium alloy in China has become a preferred alloy for the structural materials in aerospace and military fields [3]. FSW process consist of rotating tool and tool pin which generates frictional heat throughout the process perhaps the weld joints are formed, the joints deform plasticity in metals this is due to the movement of tool at the joining edges. Fig .1 shows welding process principle and the major process parameter mainly rotational speed, tool geometry, axile downward force, frictional temperature which are involved in development of quality joints [3, 4] as shown in Fig. 1.

Content from this work may be used under the terms of the Creative Commons Attribution 3.0 license. Any further distribution of this work must maintain attribution to the author(s) and the title of the work, journal citation and DOI. Published under license by Materials Research Forum LLC.

Materials Joining and Manufacturing Processes: MJMP 2025   Materials Research Forum LLC
Materials Research Proceedings 55 (2025) 102-109   https://doi.org/10.21741/9781644903612-16

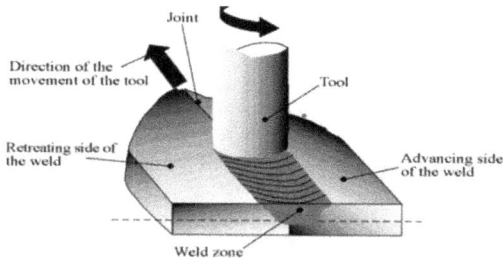

Fig. 1 FSW principle [5]

The FSW joints contain mainly four major zones namely stir zone, heat affected zone, thermomechanical affected zone, and the base metal as shown in Fig. 2, compared to all the zones stir zone exhibit fine and equisized grains. In recent study many researchers designed joints using FSW process and reported better quality of joints [5]. In depth study is required to have a proper knowledge on process and the mechanical, microstructural characteristics of FSW weld joints sections, as per the thickness of the workpiece considered in FSW process is capable up to 10 mm thickness.

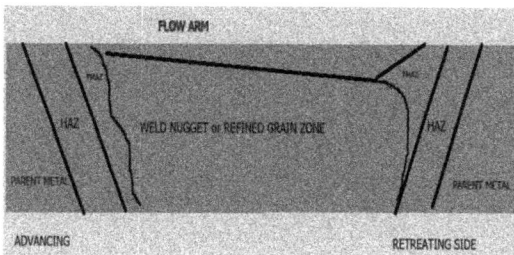

Fig. 2 FSW joint cross section with major zones [6].

Further, Gupta's et al. [6] reported that for application in aerospace industries many researchers looked up for the 6xxx, 7xxx series and the mechanical behaviour of developed aluminium alloys. Review investigates for the engineering applications the cast and weight of metals are primary considered and preferred most, also the mechanical properties such as tensile strength and the hardness significantly increase with decrease in impact strength. Rahul et al. [7] have reported the shoulder move in weld direction and shoulder pin result in dynamic recrystallization presenting in fine grain formation. Omar S. Salih et al [8] reported that FSW is effective for join aluminium alloys which are capable to crack in solid and liquid form at the weld zone and heat affected zone respectively.

Various tool pins of FSW process:
In FSW tool do major role in welding process, it has some unique quality that forms plastic flow in which it makes formation of fine and high strength weld joint. FSW tools are of different shapes normally conical, pedal, thread, cylindrical, triangular, square, and pentagonal, as mentioned have different textures, where square tool is best in all and produces 78% of efficiency, square tool makes homogeneous distribution this is due to no vertical movement of material, also for obeying of high eccentricity, produces high temperature as compared to many other tool pin profiles

produce better weld flow [9].Fig - 3 displays different tool pin profile like cylindrical, triangular, square, trapezoidal of different dimensions that mix and weld the workpiece material, as effective welding considered a triangular tool pin profile is better than the square tool pin profile. [10].

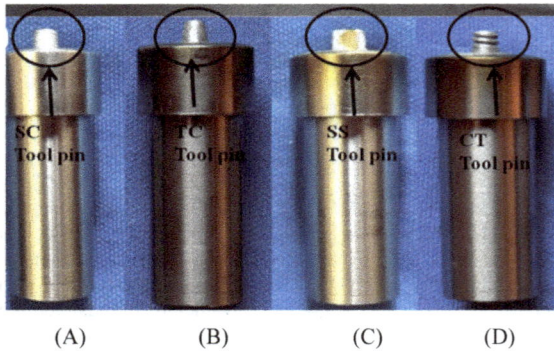

Fig. 3 *Different type of FSW tool pins (A) Cylindrical (B) Triangular (C) Square (D) Trapezoidal [10]*

Nidhi et al. [11] reported move of material from front to back of tool which is directly depends on the statical volume to swept volume ratio (ST/SW) square tool pin profile has maximum ST / SW ratio compared to cylindrical, triangular, trapezoidal and also give better weld action. Biradar et al. [12] reported insufficient heat input to joints that cause defects, for the taper tool pin defect in form of voids this is due to lower weld strength of joints, in cylindrical tool pin have higher weld strength of 37%, then the taper tool pin.

Mechanical Properties of FSW joints:

Tensile Properties

Many authors have given knowledge on the tensile properties of FSW joints. As shown in Fig-4 the ultimate tensile strength for the parent cell 455 MPa and FSW weld specimen 320 MPa the apparent strain value around 22% for both the cases respectfully. In the HAZ region ductile fracture, rough and high strength precipitates are formed [13]. Patil et al. [14] as reported that AA8081 B_4C the ultimate tensile strength value is 220 MPa at 108.3HV for maximum which improves weight at 6%. Chandrasekar et al [15] reported that by conduction of mechanical test AA8011 TiC, WC, B_4C, ZrC found that ultimate tensile strength 430 MPa respectfully. review study shows mechanical properties of weld joints are optimized and systematically examined. Rashad M et all. [16] reported that ultimate tensile strength AA8009/ AA8011 is 110 MPa, as conduction of heat treatment penning is used to strengthen joints and to get ultra fine grains. Ashu et al. [17] reported that compared to other tool pin profiles triangular pin profile have high tensile strength and low tensile strength for cylindrical tool pin profile for the of FSW joints.

Materials Joining and Manufacturing Processes: MJMP 2025   Materials Research Forum LLC
Materials Research Proceedings 55 (2025) 102-109   https://doi.org/10.21741/9781644903612-16

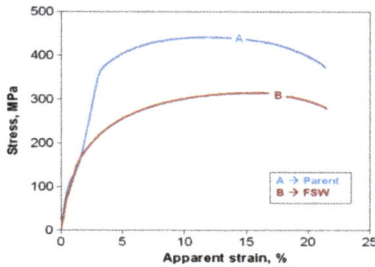

Fig. 4 *Stress vs strain curve of friction stir weld joint [18].*

Microhardness properties

The microhardness of the AA8009 FSW joints intensify of microhardness of joint will increase with increase traverse speed, if more the traverse speed the reduction in the hardness. 90 HV is minimum hardness value found for FSW joint with the welding speed of 40 mm/min, which is because of huge supply of heat to the weld length, where the hardness of weld joint is directly related to temperature distribution at weld joints. As shown in Fig. 5 microhardness value at the weld region is less than the base metal, reported that mechanism of solid strengthening occurs in welding condition and referred as primary mechanism. At age treating condition strain localisation occurs due to distribution of particles. [19].

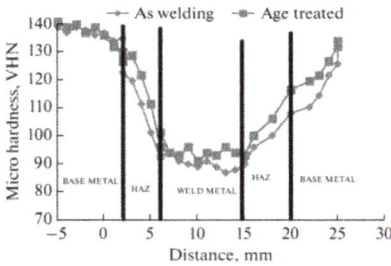

Fig. 5 *Microhardness variation for different traverse speed [19].*

In review of AA8xxx FSW joints, inside nugget zone there is an increase in the natural aging of microhardness, also composition of joint precipitation will help in the improvisation of hardness [20].

Materials Joining and Manufacturing Processes: MJMP 2025
Materials Research Forum LLC
Materials Research Proceedings 55 (2025) 102-109
https://doi.org/10.21741/9781644903612-16

Table 1 *Hardness and tensile stress of various AA8xxx joint obtained by Friction stir welding process.*

S. No	MATERIAL	PROCESS PARAMETERS	UTS (MPa)	HARDNESS (HV)	REMARKS
1.	AA8081 A-H14 [21].	Frictional heat 550^{0}c Square tool pin, Rotational speed 1600rpm.	223.5	108.3	Observed that both values are increa when impact strengtl decreased and produ better quality of w joints.
2.	AA8011 [22].	Tilt angle, tool pin dimensions Rotational speed 1600rpm	100.12	107.6	Observed high h generation which cau turbulence grain gro in stir zone.
3.	AA8011- h14 and AA6061-T6 [23].	Downward force, tilt angle, Welding speed 50mm/s.	77.88	70.17	Observed value do depend on tool offset location of alloy, so alloy placed in adva side.
4.	AA8006 - B$_4$C [24].	Frictional temperature 400°c rotational speed, 3mm pin dimensions.	220	205.4	Observed that strengthening enhance the strength weld joints.
5.	AA8011- AA7475 [25].	Square tool pin, Weld speed 65mm/min, 1200rpm rotational speed Downward force.	180	67.27	Observed improvem in the values of ten strength with increase welding and rotatic speed.
6.	AA8009 [26].	Rotational speed 1200 rpm, tilt angle +or-5 Frictional temperature	440	110	Observed that high strain, strain rate dur FSW and formation shear bonding exhibited apprecia softening.

Microstructural characterization

Most of researchers have given report on microstructure characteristics of weld joints produced by Friction stir welding process. The nugget zone (NZ), heat affected zone (HAZ), thermo mechanically affected zone (TMAZ), and base metal (BM) are the four major microstructure zones which are formed in welding process. as shown Fig - 6. development and evolution of microstructure have been studied and discussed.

Materials Joining and Manufacturing Processes: MJMP 2025
Materials Research Proceedings 55 (2025) 102-109

Materials Research Forum LLC
https://doi.org/10.21741/9781644903612-16

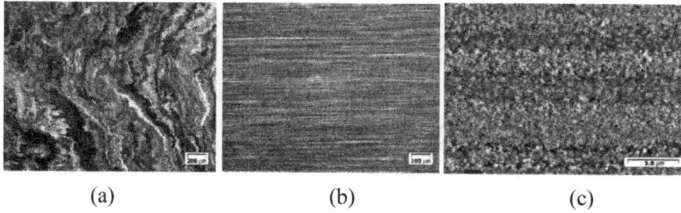

(a)   (b)   (c)

Fig. 6 *Microstructure zones of AA8xxx FSW joint (a) cross section at centre with extrusion direction is vertical direction, (b, c) transverse section at perpendicular to extrusion direction.*

Fig. 6 a, b displays the view of base metal at centre of plate thickness, Fig - 6 c shows the variation of size and density inside base metal microstructure. The light micrographs are (a, b) and SEM/BSE (c) micrographs of AA-8009 base metal, where in $Al_{13}(Fe\ V)_3$ Si is distributed in range of 50 to 100 nm diameter, found inside and outside boundaries of the fine, these fine grains of alpha aluminium are ranging from the 0.5 to 1 micron in diameter [26].

Huang et al. [27] reported and described the dissimilarities of FSW joints, also made important study on grain refining process and the parameters of the nugget zone (NZ). Where FSW consist of two major sides namely advancing side (AS) and retreating side (RS). AS is the side in which direction of tool rotation is in same direction of the welding, the RS is side in which direction of tool rotation is reverse to the direction of welding. Kumar S et al. [28] reported nugget zone show fine grains these grains are procured without pre heating the alloys. Rajath Kumar Roy et al. [29] reported that in AA8011 and AA8009 series weld joints contains same sized grains, where it is due to the dynamic recrystallization and the grains are recrystalised in nugget zone. The rotational tool speed about 900 RPM with speed of welding about 40 mm/min, a fine and uniformly distributed grains are formed in the weld region this will manifest higher tensile strength of these respective joints.

Conclusions

This review study, concentrate on AA8xxx series weld joints using friction stir welding and obtained following conclusions:

i.   FSW process parameters like rotational speed, tilt angle, traverse speed which affects the quality of weld joints.

ii.   There are four main microstructural zone, namely base material, nugget zone, HAZ, TMAZ. Fine grains are formed in TMAZ and NZ and they are equally dispersed.

iii.   Joints of AA8xxx have better ductility and tensile strength. However, joints undergo failure in HAZ zone.

iv.   Most of authors reported that square tool pin produces the better joint efficiency, compared to other tool pins.

In addition, treatments method like heats and cold to be carried out for weld joints to analyse the FSW joint strength characteristics. Also, artificial intelligence (AI) could be employed to detect weld defects and enhance the joint efficiency.

Acknowledgement

The Authors Gratefully Acknowledge the REVA University Bengaluru for the support of funding and providing the facility to carry out the Research work Ref No: RU/SEED/MECH/2025/44

References

[1] S. Govinda raj, Karthikeyan, Investigation on mechanical properties welded aluminium joints of AA8011 using Friction Stir Welding. Int J of A. Eng Res 10 (2015). 0973-4562. http://www.ripublication.com.

[2] H. W. Zhang, Zhao Zhang, and J. T. Chen. The finite element simulation of the friction stir welding process. Mat Sci and Eng 2 (2005). 340-348. https://doi.org/10.1179/174329307X177919

[3] S. Kailainathan. et al. Influence of Friction Stir Welding Parameter on Mechanical Properties in Dissimilar (AA6063-AA8011) Aluminium Alloys. Int J of Inn Res in Science, Engineering and Tech 3 (2014). 15691-15695. https://doi.org/10.1016/j.jmapro.2018.12.014

[4] Zhang, Chenghang, et al. On the microstructure and mechanical properties of similar and dissimilar AA7075 and AA2024 friction stir welding joints: Effect of rotational speed. J of Manf Pro 37 (2019). 470-487. https://doi.org/10.1016/j.jmapro.2018.12.014

[5] Sachinkumar, S. Narendranath, and D. Chakradhar. Microstructure, hardness and tensile properties of friction stir welded aluminum matrix composite reinforced with SiC and fly ash. Silicon 11 (2019). 2557-2565. https://link.springer.com/article/10.1007/s12633-018-0044-5

[6] Gupta, Vivudh, Balbir Singh, and R. K. Mishra. Microstructural and mechanical characterization of novel AA7075 composites reinforced with rice husk ash and carbonized eggshells, Inst of Mech Eng, Part L J of Mat Dign and App 235.12 (2021). 2666-2680. https://doi.org/10.1177/14644207211031265

[7] B. Rahul, and Sachinkumar Patil. Tensile, microhardness and microstructural characteristics of friction stir welded/processed AA7075 alloy: A review. Mat T Proc (2023). https://doi.org/10.1016/j.matpr.2023.07.299

[8] Salih, S. Omar. et al. A review of friction stir welding of aluminium matrix composites. Materials & Design 86 (2015) 61-71. https://doi.org/10.1016/j.matdes.2015.07.071

[9] Xu, Yang, et al. Precipitation behavior of intermetallic compounds and their effect on mechanical properties of thick plate friction stir welded Al/Mg joint. J of Manf Pro 64 (2021). 1059-1069. https://doi.org/10.1016/j.jmapro.2020.12.068

[10] Sachinkumar, S. Narendranath, and D. Chakradhar. Characterization and evaluation of joint properties of FSWed AA6061/SiC/FA hybrid AMCs using different tool Pin profiles. Tran of I I M 73 (2020). 2269-2279. https://link.springer.com/article/10.1007/s12666-020-02035-2

[11] Sharma, Nidhi, et al. "Material stirring during FSW of Al–Cu: Effect of pin profile. Mat and Manf Pro 33.7 (2018). 786-794. https://www.tandfonline.com/doi/abs/10.1080/10426914.2017.1388526

[12] B. Rahul, and Sachinkumar Patil. A Systematic Review on Microhardness, Tensile, Wear, and Microstructural Properties of Aluminum Matrix Composite Joints Obtained by Friction Stir Welding: Past, Present and Its Future." Tran of the I I M (2024). 1-15. https://link.springer.com/article/10.1007/s12666-024-03303-1

[13] Sharma, Rajesh, M. K. Pradhan, and Pankaj Jain. Fabrication, characterization and optimal selection of aluminium alloy 8011 composites reinforced with B 4 C-aloe vera ash. Mat Res Exp 10.11 (2023). 116513-16. https://iopscience.iop.org/article/10.1088/2053-1591/acec32/meta

[14] Patil, Kumar Lingaraj, Mousil Ali, and Madeva Nagaral. Studies on al8081-b4c metal matrix composites fabricated by stir casting method." Int J M Eng Res (2014). IJMER 4-26.https://d1wqtxts1xzle7.cloudfront.net/34746607/IJMER-47060104

[15] P. Chandrasekar, S. Natarajan, and K. R. Ramkumar. Influence of carbide reinforcements on accumulative roll bonded Al 8011 composites. Mat and Manf Pro 34.8 (2019). 889-897. https://doi.org/10.1080/10426914.2019.1594279

[16] R. Muhammad, et al. Effect of alumina and silicon carbide hybrid reinforcements on tensile, compressive and microhardness behaviour of Mg–3Al–1Zn alloy. Materials Characterization 106 (2015). 382-389. https://doi.org/10.1016/j.matchar.2015.06.033

[17] Garg, Ashu, Madhav Raturi, and Anirban Bhattacharya. "Strength, failure and microstructure development for friction stir welded AA6061-T6 joints with different tool pin profiles." CIRP J of Manf Sci and Tech 29 (2020). 99-114. https://doi.org/10.1016/j.cirpj.2020.03.001

[18] P. Srinivasan, Bala, et al. Characterisation of microstructure, mechanical properties and corrosion behaviour of an AA2219 friction stir weldment. Journal of Alloys and Compounds 492.1 (2010). 631-637. https://doi.org/10.1016/j.jallcom.2009.11.198

[19] S. Arunkumar. et al. Microstructural and mechanical characterization of as weld and aged conditions of AA2219 aluminium alloy by gas tungsten arc welding process. Rn J of Non-Ferr Met 59 (2018). 93-101. https://link.springer.com/article/10.3103/S1067821218010030

[20] C. Elanchezhian. et al. Parameter optimization of friction stir welding of AA8011-6062 using mathematical method. Pro Eng 97 (2014). 775-782. https://doi.org/10.1016/j.proeng.2014.12.308

[21] Engler, Olaf, Johannes Aegerter, and Dirk Calmer. Control of texture and earing in aluminium alloy AA 8011A-H14 closure stock. Mat Sci and Eng A 775 (2020). 138965. https://doi.org/10.1016/j.msea.2020.138965

[22] Davidson, B. Shiloh, and S. Neelakrishnan. Influence of friction stir welding parameters on tensile properties of AA8011 aluminium alloy plate. J of Compt and Thry of N sci 15.1 (2018). 939822.   https://doi.org/10.1166/jctn.2018.7060

[23] K. Palani. et al. Hybrid Fuzzy based response surface optimization of welding parameters on Vickers microhardness and impact strength of FSWeld AA8011-H24 aluminium alloy joints. Mat T Proc 23 (2020). 573-582. https://doi.org/10.1016/j.matpr.2019.05.412

[24] F. Khodabakhshi, A. P. Gerlich, and M. Worswick. Fabrication and characterization of a high strength ultra-fine-grained metal-matrix AA8006-B4C layered nanocomposite by a novel accumulative fold-forging (AFF) process. Mat & Dign 157 (2018). 211-226. https://doi.org/10.1016/j.matdes.2018.07.047

[25] S. Dharmalingam, and K. Lenin. Effect of friction stir welding parameters on microstructure and mechanical properties of the dissimilar (AA7475-AA8011) joints. Mat T Proc 39 (2021). 105-109. https://doi.org/10.1016/j.matpr.2020.06.317

[26] R. Arun Kumar. et al. Examination of the mechanical, corrosion, and tribological behavior of friction stir welded aluminum alloy aa8011. Tran on Mar Sci 10.01 (2021). 20-41. https://hrcak.srce.hr/258057

[27] Hu, Z. L. et al. The effect of postprocessing on tensile property and microstructure evolution of friction stir welding aluminum alloy joint. Mat Char 99 (2015). 180-187. https://doi.org/10.1016/j.matchar.2014.11.015

[28] R. Kumar. et al. "Effects of tool pin profile on the formation of friction stir processing zone in AA1100 aluminium alloy." Mat T Proc 48 (2022). 1594-1603. https://doi.org/10.1016/j.matpr.2021.09.491

[29] R. Rajat Kumar, Sujoy Kar, and D. Siddhartha. Evolution of microstructure and mechanical properties during annealing of cold-rolled AA8011 alloy J of Al and Comp 468.1 (2009). 122-129. https://doi.org/10.1016/j.jallcom.2008.01.041

Materials Joining and Manufacturing Processes: MJMP 2025
Materials Research Proceedings 55 (2025) 110-117

Materials Research Forum LLC
https://doi.org/10.21741/9781644903612-17

Structural evaluation of microwave butt welded MONEL 400 sheets through finite element analysis

Gajanan M. Naik[1,a], Shanthala K.[1,b], Sadashiv Bellubbi[2,c], Chirag L.[1,d], Devendra Gowda[3,e], Suresh Poyil Subramanyam[4,f*]

[1]Department of Mechanical Engineering, RV Institute of Technology and Management, JP-Nagar-560076, Bengaluru, Affiliated to Visvesvaraya Technological University, Belagavi. Karnataka, India

[2]Department of Mechanical Engineering, Jain College of Engineering and Research, Udyambag-590008, Belagavi, Affiliated to Visvesvaraya Technological University, Belagavi. Karnataka, India

[3]Department of Electrical and Electronics Engineering, Acharya Institute of Technology, Bengaluru, Affiliated to Visvesvaraya Technological University, Belagavi. Karnataka, India

[4]Department of Mechanical Engineering, Alva's Institute of Engineering and Technology, Moodbidri, Affiliated to Visvesvaraya Technological University, Belagavi. Karnataka, India

[a]gajamnaik@gmail.com, [b]kshanthala2@gmail.com, [c]bellubbisadashiv@gmail.com, [d]chirag6472@gmail.com, [e]devendragowdapatil@gmail.com, [f]pssuryatech@gmail.com

Keywords: Microwave Welding, Monel-400, Thermal Analysis, Structural Analysis, ANSYS

Abstract. Microwave welding is a method of joining materials that makes use of microwave energy for accurate and efficient bonding. Unlike conventional welding methods reliant on electrical arcs or flames for heat generation, microwave welding employs electromagnetic waves within the microwave frequency spectrum to induce targeted heating and fusion at designated bonding sites. This study investigates the structural integrity of a butt-welded joint of Monel-400 a nickel-copper alloy, with nickel- powder as interlayer in a microwave welding setup. This study provides the static structural analysis of a beam via ANSYS 2024 R1 software. Three important stress distributions are illustrated: maximum principal stress, minimum principal stress, and total deformation. The maximum principal stress shows critical stress points that can compromise structural integrity, while the minimum principal stress determines areas of failure. The deformation visualization provides an understanding of the material's response to applied loads, guaranteeing design reliability. This research highlights the significance of computational analysis in engineering design and safety ratings.

Introduction

Microwave welding is a welding process that uses electromagnetic radiation in the microwave frequency range to heat and join materials without direct contact. It offers several advantages compared to traditional welding methods. The key advantage is the ability to generate heat directly within the material being welded, leading to faster and more precise heating. This feature allows for high-quality welds with minimal distortion and limited thermal damage to the surrounding areas. As a result, microwave welding has found applications in diverse industries such as automotive, aerospace, electronics, and medical devices [1].

In Microwave heating, microwaves interact with molecules, and heat is generated inside the material due to molecular agitation. The interaction of microwaves with bulk metals is poor at room temperature due to low skin depth. To overcome this problem, Microwave Hybrid Heating (MHH) is used along with the use of metallic powders, which are good absorbers of microwaves [2]. MHH

Content from this work may be used under the terms of the Creative Commons Attribution 3.0 license. Any further distribution of this work must maintain attribution to the author(s) and the title of the work, journal citation and DOI. Published under license by Materials Research Forum LLC.

Materials Joining and Manufacturing Processes: MJMP 2025   Materials Research Forum LLC
Materials Research Proceedings 55 (2025) 110-117   https://doi.org/10.21741/9781644903612-17

combines microwave heating with conventional heating, in which microwaves couple with a material called susceptor and are absorbed by it.

Most of the studies that have been conducted on this subject are limited to experimental analysis and post-joint characterization. However, certain results such as the distortion and distribution of electric field, distribution of temperature, and thermal stress across the joint cannot be experimentally determined. These results are very useful in the design of microwave hybrid heating processes to achieve efficient joining and can be obtained through modelling and simulation [3]. The joining of stainless steel 316L and copper using MHH was simulated in COMSOL Multiphysics software, in which Cu-powder was used as an interface layer, and it was found that on increasing the exposure time, the porosity in the interlayer decreased [4]. The simulation and experimental analysis of Microwave hybrid sintering of Aluminium were performed and the results were compared. Simulation results agreed with the experiments and the error margin was found to be within 10 % [5]. Microwave joining of SS304 samples was studied using two different interface powders, Inconel 718 and Nickel powder. A simulation of selective microwave hybrid heating was performed and then a comparison was drawn to determine the better interface material [3]. Simulation and experimental study were performed for MS pipe joining using selective microwave hybrid heating with Nickel interfacial powder. It was found that due to the use of a carbon-based susceptor (charcoal), carbide precipitation took place in the joint zone [6]. Bellubbi et al. [7] proposed method to makes use of single-V welding type configuration to produce drum shells with ceramic backing strip prearrangement.

For the present study, Monel-400 nickel-copper alloy was selected as the primary material for the workpiece due to its favourable mechanical properties and corrosion resistance. To facilitate the joining process, nickel-interlayer powder was chosen for its exceptional characteristics, especially its corrosion resistance and mechanical strength at elevated temperatures. Using Ansys 2024R1 software, a simulation study was conducted to investigate structural integrity of butt welding. The simulation focused on analysing structural integrity of joint under temperature load and structural load across the joint region. This comprehensive analysis aimed to provide insights into the behaviour and performance of the joined materials under microwave heating conditions, thereby contributing to the optimization of the joining process and ensuring the integrity of the resulting joint.

Experimental Work

The experimental procedure for microwave butt welding of sheets of MONEL 400 was conducted using a Samsung ME83D 23L home microwave oven, rated at 2.45 GHz and 800 W maximum output power. $50\times20\times1$ mm3 sheets of MONEL 400 were pretreated with acetone to sweep away surface residue. To increase the microwave absorption, a silicon carbide (SiC) susceptor was put at the sheet interface to create localized heating within the weld zone. The sheets were positioned in a butt assembly inside the microwave chamber, with a controlled compressive force introduced by a ceramic weight to have uniform contact. Microwave exposure time was optimized between 120 and 300 seconds, according to initial trials, for sufficient fusion with minimal thermal degradation. Temperature monitoring was performed with a Fluke 572-2 infrared thermometer to ensure that the weld zone was about 900–1100 °C, which is adequate for metallurgical bonding. The microwave welding setup is shown in Figure 1.

Materials Joining and Manufacturing Processes: MJMP 2025

Materials Research Forum LLC

Materials Research Proceedings 55 (2025) 110-117

https://doi.org/10.21741/9781644903612-17

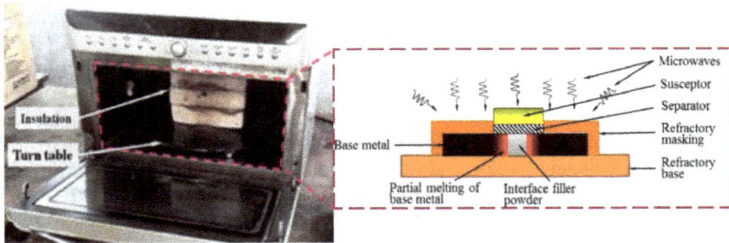

Fig. 1 Microwave welding set up

Modelling and Simulation

The welded part was first modeled in SolidWorks, where its geometric and dimensional requirements were modeled with care shown in Figure 2 (a). For compatibility with ANSYS, the model was exported in IGES format so that it could be imported into the simulation environment without any issues. This allowed for precise representation of the welded joint so that the finite element analysis (FEA) process could capture the required structural and thermal behavior. The imported model in ANSYS was then ready for simulation by specifying the weld interface and the heat-affected zones to precisely represent the welding process.

Material properties were assigned to simulate those of MONEL 400, including important parameters like a Young's Modulus of 179 GPa, a Poisson's Ratio of 0.32, and a Density of 8.83 g/cm³. In order to achieve thermal effects, a thermal conductivity of 21 W/(m·K) was specified to ensure correct heat transfer simulation. The part was subjected to structural and thermal loads in the form of a static load of 1000 N and a thermal load of 50°C, representative of actual operating conditions. Also, boundary conditions were defined to replicate actual constraints like fixed supports on some points and prescribed temperatures on others to closely simulate the welded structure under anticipated service conditions.

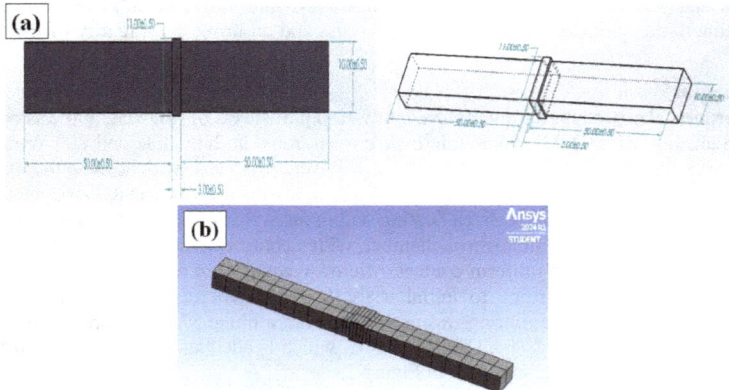

Fig. 2. a) Modelling of butt welding with interface layer b) Meshing

A systematic meshing strategy was followed (Fig. 2(b)). The model employed a tetrahedral mesh in order to effectively capture complicated geometries as well as   carry   out   accurate analysis of stress distribution, setting the initial mesh size to 1 mm and subsequently reducing it to 1.4 mm to achieve solution convergence and efficient computation. Simulation data were

Materials Joining and Manufacturing Processes: MJMP 2025   Materials Research Forum LLC
Materials Research Proceedings 55 (2025) 110-117   https://doi.org/10.21741/9781644903612-17

evaluated with respect to key mechanical properties, such as yield strength (350 MPa) and elastic characteristics (Young's Modulus of 179 GPa). Heat distribution throughout the welded joint was also investigated for heat dissipation and possible thermal stresses. This data gave informative insight into welded joint structural integrity and durability and informed possible design optimizations and processing improvements for high-performance applications.

Results and Discussion

Thermal Analysis

The thermal analysis presented here (Fig 3a) shows significant features of heat distribution and flux in a given model through the ANSYS software. The visualizations are divided into three sub-figures 3 (a), (b), and (c), representing various parameters of thermal study: total heat flux, directional heat flux, and temperature distribution, respectively. The comprehension of these thermal attributes is an imperative for applications where thermal management plays a critical role, for example, in materials science, mechanical engineering, and electronics cooling.

i) Total Heat Flux Analysis: Figure 3 (a), the total heat flux is illustrated in watts per meter squared (W/m^2), indicating heat flow over the model. Color gradient illustrates changes in heat distribution, where regions indicate highest total heat flux around $1.11316 \ MW/m^2$. High heat flux here indicates heat concentration or heat transfer, perhaps by particular boundary conditions or material settings. The color map varies from blue, cooler shades to red, hotter shades, enabling an intuitive visual recognition of thermal gradients that may affect material performance and system stability.

ii) Directional Heat Flux: Figure 3 (b) shows directional heat flux along the x-axis, where the value reaches a maximum of $1.12926 \ MW/m^2$. This specifies the particular direction from which the heat is flowing, a necessary understanding of thermal paths in ensuring that materials have the ability to dissipate heat properly. The visualization can determine the area where heat tends to accumulate, which requires additional design considerations to prevent overheating.

iii) Temperature Distribution: Figure 3 (c) illustrates temperature distribution across the model in Kelvin with the highest recorded temperature at 1300 K. Temperature mapping is important for determining thermal material limits applied in the model. Examination of such temperature gradients is useful for identifying regions that can benefit from thermal insulation or more efficient heat dissipation methods. Together, the information derived from the thermal analysis guides design and operational procedures for efficient thermal management.

Materials Joining and Manufacturing Processes: MJMP 2025   Materials Research Forum LLC
Materials Research Proceedings 55 (2025) 110-117   https://doi.org/10.21741/9781644903612-17

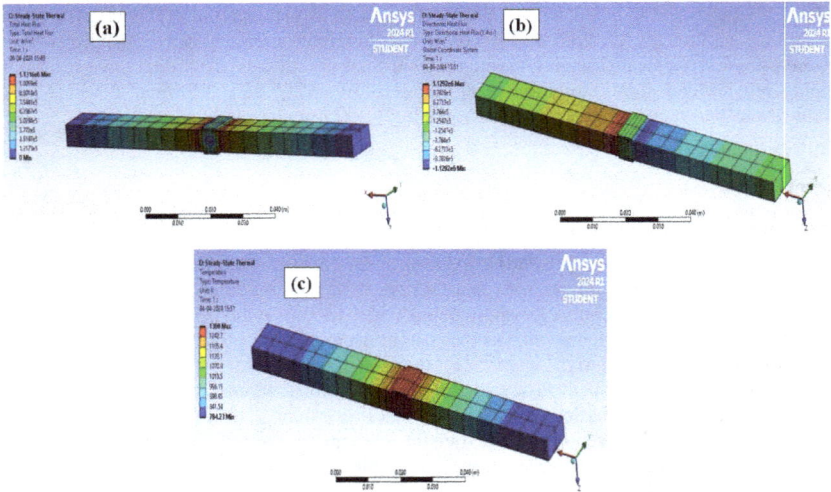

Fig.3 a) Total Heat Flux b) Directional Heat Flux c) Temperature distribution

Static Structural Analysis

Figure 4 represents the outcome of a static structural analysis conducted using ANSYS software, which shows the maximum and minimum principal elastic strains that an object undergoes under loading conditions. In figure 4 (a), the maximum principal elastic strain is shown, where regions of maximum deformation of the material under stress are indicated. The color gradient is from blue (minimum strain) to red (maximum strain), with the maximum strain value at around (1.122 10^-2). This indicates that specific parts of the structure are strained highly, which might be important for evaluating possible failure modes or regions that need to be reinforced.

The figure 4 (b) shows the minimum principal elastic strain, which reveals areas that are subjected to compressive stress. Like the visualization of maximum strain, this figure uses a color scale to represent strain values, with red hues representing high compressive strain and blue representing low values. The highest compressive strain measured is approximately (-1.342 10^5), highlighting regions of the structure that may be prone to buckling or other types of mechanical failure. Through examination of the distribution of these values in both figures, engineers can make informed design changes, material choices, and overall structural integrity evaluations to guarantee safety and performance under operating conditions.

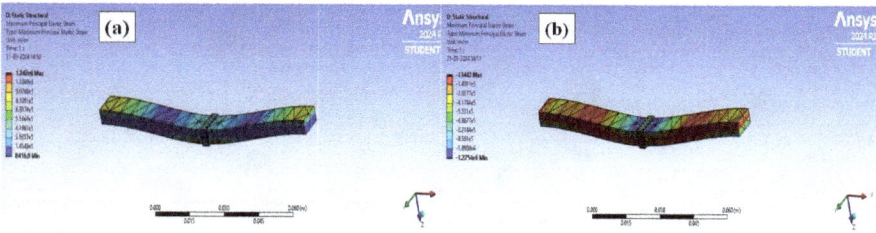

Fig. 4 a) Maximum Principal Strain b) Minimum Principal Strain

Figure 5 shown other structural analysis results achieved with the use of ANSYS software, each for a different type of stress and total deformation values for a given component. The three figures, designated as 5 (a), (b), and (c), show the maximum principal stress analysis, minimum principal stress analysis, and total deformation analysis, respectively. Each image displays a visualization of stress distribution and deformation behavior under static loading conditions.

i) Principal Stress Analysis: In figure (a), the Maximum Principal Stress is shown. This is the analysis that shows the maximum stress values the structure is subjected to, presented through a color-coded scale from green (minimum stresses) to red (maximum stresses). The highest principal stress observed is about 2.34 MPa, and this indicates that some parts of the structure are subjected to excessive loading. These points of stress concentration are significant since they may cause failure if they reach beyond the material's yield point. The data indicates that points near the center of the ellipse have the most stress, thus the need for emphasis on them in case design can be made better.

ii) Minimum Principal Stress: Figure 5 (b) gives information about the Minimum Principal Stress, with the color gradient here also pointing towards compressive stresses. The value of the minimum principal stress is approximately 9.47 MPa. Compressive stresses may represent buckling or instability in the structural members. It is crucial to analyze these values along with the maximum stress data to check whether the material can sustain the anticipated loads under tension and compression without sustaining any damage.

iii) Total Deformation Analysis: Figure 5 (c), the Total Deformation analysis is shown. This analysis calculates the extent to which the structure deforms when subjected to applied loads, with the maximum recorded deformation being around 2.22 x 10^-5 m. The color mapping shows different degrees of deformation, with regions having greater shifts prominently highlighted. Total deformation knowledge is essential as it dictates structural integrity and working effectiveness; inordinate deformation could impair functional functionality or endanger safety. The observations from this photograph along with the stress analyses help create an end-to-end awareness of the structural attributes, facilitating enhancements or refinements in design.

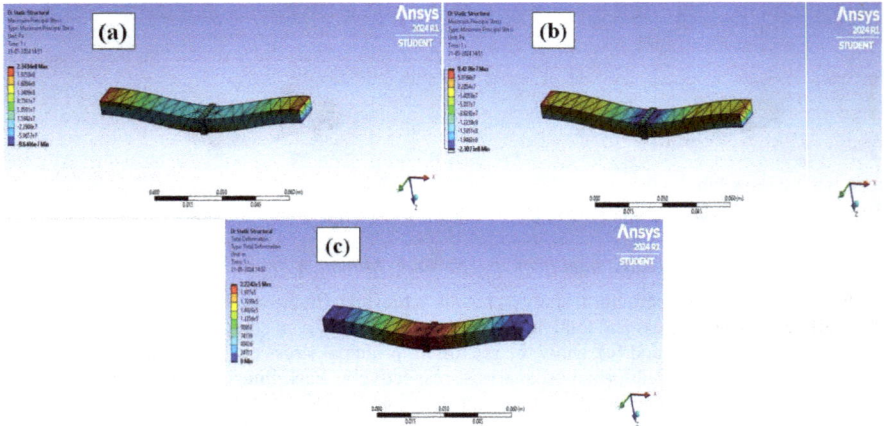

Fig.5 a) Maximum Principal Stress b) Minimum Principal Stress c) Total Deformation

Conclusion

The study focused on assessing the structural integrity of a butt-welding of Monel-400 nickel-copper joined through a nickel-interlayer powder via microwave welding. Through simulation studies conducted using Ansys 2024R1 software, effect of thermal load and structural load across the joint region were analysed and following conclusions are drawn.

- The thermal analysis revealed significant findings regarding heat flux, directional heat flux, and temperature distribution, providing insights into the thermal behaviour of the materials under microwave heating conditions.
- The output shows the maximum principal stress distribution over the structure with a maximum value of around 2.34 MPa. This identifies regions of failure or yielding potential under load.
- The minimum principal stress values reflect compressive stress levels, which have a maximum of around 9.47 MPa. This assists in developing an understanding of the structural robustness and resilience of the component under working loads.
- The accumulated deformation values, with the largest being about 2.22 mm, indicate that the structure undergoes heavy displacement, which could influence its performance under function and life of service.

References

[1] E. Goswami, A. K. Srivastava, A. Kumar, M. Salman, M. Z. Choudhary, and S. S. Pandey, Review of Microstructural and Mechanical Properties of Microwave Welding of Lightweight Alloys, Model. Charact. Proc. Smart Mater. 1 (2023) 185-204. https://doi.org/10.4018/978-1-6684-9224-6.ch009

[2] S. Rawat, R. Samyal, R. Bedi, and A. K. Bagha, A comparative study of interface material through selective microwave hybrid heating for joining metal plates, Mater. Today: Proc. 65 (2022) 3117-3125. https://doi.org/10.1016/j.matpr.2022.05.346

[3] K. V. B. Reddy, K.V. Hari Shankar, G. Venkatesh, and R. R. Mishra, Numerical simulation study on microwave joining of Hastelloy C-276 plates using Inconel-718 interface powder, Mater. Today: Proc. 98 (2024) 187- 193. https://doi.org/10.1016/j.matpr.2023.10.104

[4] S. Tamang, and S. Aravindan, Joining of dissimilar metals by microwave hybrid heating: 3D numerical simulation and experiment, Int. J. Therm. Sci. 172 (2022) 107281. https://doi.org/10.1016/j.ijthermalsci.2021.107281

[5] N. K. Bhoi, D. K. Patel, H. Singh, S. Pratap, and P. K. Jain, Multi- physics simulation study of microwave hybrid sintering of aluminium and mechanical characteristics, Proc. Inst. Mech. Eng., Part E 236 (2022) 1989-1996. https://doi.org/10.1177/09544089221074829

[6] M. Pal, V. Kumar, S. Sehgal, H. Kumar, K. K. Saxena, and A. K. Bagha, Microwave hybrid heating based optimized joining of SS304/SS316, Mater. Manuf. Processes 36 (2021) 1554-1560. https://doi.org/10.1080/10426914.2020.1854469

[7] S. Bellubbi, and N. Sathisha, Reduction in through put time of drum shell manufacturing by single-V welding configuration, IOP Conf. Series: Mater. Sci. Eng. 376 (2018) 012104. https://doi.org/10.1088/1757-899X/376/1/012104

Materials Joining and Manufacturing Processes: MJMP 2025 Materials Research Forum LLC
Materials Research Proceedings 55 (2025) 118-129 https://doi.org/10.21741/9781644903612-18

Optimization of mechanical properties in 3D-printed PLA parts with honeycomb and cubic infill patterns using taguchi method

Aveen K.P.[1,a]*, Shivaramu H.T.[1,b] Praveen K.C.[2,c], Sridhar D.R.[1,d] and Thoran[1]

[1]Department of MechanicaL Engineering, Mangalore Institute of Technology And Engineering, Karnataka, Moodabidri- 574225, Karnataka India

[2]Department of MechanicaL Engineering, Alva's Institute of Engineering & Technology, Karnataka, Moodabidri- 574225, Karnataka India

[a]aveen@mite.ac.in, [b]shivaramu@mite.ac.in, [c]praveenkc@aiet.org,in, [d]sridhar@mite.ac.in

Keywords: 3D Printing, FDM, PLA, Taguchi Optimization, Cubic Infill, Honeycomb Infill

Abstract. 3D printing is a revolutionary manufacturing technology that provides previously unheard-of levels of efficiency and design freedom. Fused Deposition Modeling (FDM) is a popular method for creating complicated parts using thermoplastic materials like Polylactic Acid (PLA). Layer height, layup speed, and infill pattern are some of the process variables that have a major impact on the mechanical characteristics of PLA components that are FDM produced. The goal of this study was to maximize the mechanical performance of PLA components made with honeycomb and cubic infill patterns. The best combinations of layer height, layup speed, and infill density were found via Taguchi optimization. The findings showed that in terms of modulus and tensile strength, honeycomb infill designs continuously performed better than cubic patterns. Notably, a 40% infill density with a 0.3 mm layer height and a 60 mm/s layup speed yielded the best mechanical properties for honeycomb-patterned parts. The Taguchi study emphasized how important layer height and infill density are to mechanical performance. These results offer useful recommendations for producers looking to enhance FDM procedures in order to create long-lasting, premium PLA components. Tensile strength and modulus were found to be most significantly impacted by infill density and layer height when the Taguchi method was used to optimize the process parameters.

Introduction

3D printing has completely changed the manufacturing industry by making it possible to create intricate parts with remarkable accuracy and little waste of material [1,2]. Because of its affordability and adaptability, Fused Deposition Modeling (FDM) is one of the most widely used 3D printing technologies [3,4]. Layers of thermoplastic materials are extruded in FDM to create three-dimensional objects. A popular thermoplastic in FDM, polylactic acid (PLA) is perfect for both functional and prototyping applications because of its biodegradability, ease of processing, and reasonably good mechanical qualities [5]. Despite its low melting temperature, moderate mechanical strength, and outstanding printability, PLA products' mechanical performance is extremely susceptible to printing parameters including infill patterns, layer height, and layup speed [6]. For PLA components to have the appropriate mechanical qualities, these parameters must be optimized. The Taguchi approach is among the best ways to optimize industrial processes, including 3D printing. Using orthogonal arrays to systematically examine the effects of several variables with a small number of tests, this reliable statistical method streamlines experimentation [7]. The final mechanical qualities of printed objects are significantly impacted by process variables like as layer height, layup speed, and infill density in the context of FDM.

The qualities of the components and the efficiency of production are greatly influenced by a number of FDM process factors. Important process parameters include things like layer thickness,

Content from this work may be used under the terms of the Creative Commons Attribution 3.0 license. Any further distribution of this work must maintain attribution to the author(s) and the title of the work, journal citation and DOI. Published under license by Materials Research Forum LLC.

raster angle, build orientation, infill density, printing speed, infill pattern, raster width, etc. [8]. The results show that reducing the layer height enhances inter-layer adhesion, leading to improved mechanical strength. However, higher lay-up speeds may reduce strength by jeopardizing the bonding between layers, as the material lacks adequate time to cool and solidify properly [9]. An important factor in the FDM process is the infill pattern, which gives the 3D print inherent support as it is built layer by layer. The effects of various infill patterns on the mechanical characteristics of 3D-printed objects have been investigated by a number of researchers. In order to evaluate their effects, Aloyaydi et al. [10] and Dezaki et al. [11] looked at patterns like grid, rectilinear, wiggle, and honeycomb. Similarly, Parab and Zaveri [12] examined the impact of infill on the compressive strength of PLA components, while Srinivasan et al. [13] examined the effects of grid, triangle, and cubic patterns on Polyethylene Terephthalate Glycol-modified [PETG] parts. Current studies demonstrate how particular patterns can improve mechanical performance, highlighting how crucial pattern selection is to getting the best outcomes. The Taguchi method has proven to be an efficient approach for optimizing these parameters, significantly reducing the number of trials needed while effectively identifying the optimal conditions for achieving maximum mechanical strength [14].

The current research focuses on analyzing the mechanical characteristics such as tensile, compressive, and flexural strength for specific infill patterns like honeycomb and cubic. While both patterns demonstrate unique performance attributes, honeycomb infill excels in stiffness-to-weight efficiency at lower densities, making it suitable for lightweight applications. On the other hand, cubic infill consistently outperforms honeycomb in terms of structural strength, particularly at higher densities, making it ideal for load-bearing applications. Despite the wealth of data on these patterns, strong statistical tools like the Taguchi approach remain underutilized in systematically optimizing the influence of process parameters on these mechanical properties.

Materials & methods
Tensile specimens were printed using PLA material in accordance with Type 1 specifications as per ASTM D638 standards, as shown in Figure 1 [15].The fabrication parameters included three different layer heights 0.15 mm, 0.2 mm, and 0.3 mm with layup speeds of 40 mm/min, 60 mm/min, and 80 mm/min. The infill densities were varied across 20%, 40%, and 60%, with cubic and honeycomb structures used as the infill patterns to evaluate mechanical performance. PLA specimens were created using a low-cost desktop 3D printer using Fused Deposition Modeling (FDM) techniques.

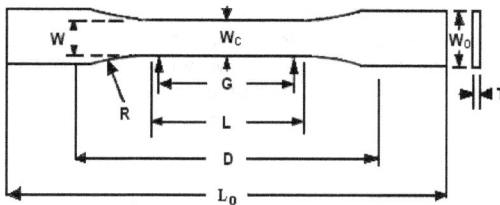

Figure 1 Tensile sample designed in 3D CAD program.

Table 1 Specimen specifications

Sl. No.	Geometrical parameter	Dimensions(mm)
1	W-width of a narrow section	13
2	L- length of a narrow section	57
3	WO-width overall, minimum	19
4	LO-length overall, minimum	165
5	G-Gauge length	50
6	D-Distance between grips	115
7	R-Radius of fillet	76

On the Kapton Tape construction surface, a total of nine pieces of each honeycomb and cubic PLA are printed. For filament extrusion, the nozzle's temperature is kept at 190^0C, and the printer platform's temperature is kept at 60^0C. Table 2 displays the printing settings, such as layer height and lay-up speed density.

Table 2: Process parameter for 3D printer.

Printing Parameter	Particulars
Layer height (mm)	0.15, 0.2 & 0.3
Lay-up speed (mm/Min)	40, 60 & 80
Infill Density, %	20, 40 & 60
Infill Pattern	Honey comb & Cubic
Nozzle dia (mm)	0.6
PLA density	1.24g/cm3

First, CAD software was used to design the tensile specimens. PrusaSlicer software was then used to slice the 3D model so it could be printed. The file was sliced, then saved in .stl format before being sent to the 3D printer to be manufactured. As shown in Figure 1, the printer was configured to generate PLA specimens with the designated infill patterns cubic and honeycomb. Every pattern was selected in order to investigate its impact on the specimens' mechanical characteristics.

One important technique for characterizing materials is the tensile test, which evaluates a specimen's strength under a uniaxial tensile force and the experimental setup is shown in figure 2. In this investigation, a universal testing machine with a 10 kN load cell was used to evaluate specimens with both honeycomb and cubic infill patterns. A controlled load was applied to each specimen at a crosshead speed of 3 mm per minute while it was positioned between the machine's grips, as shown in Figure 3. Table 2 provides a summary of the tensile strength and elongation findings from these testing.

Materials Joining and Manufacturing Processes: MJMP 2025
Materials Research Proceedings 55 (2025) 118-129

Materials Research Forum LLC
https://doi.org/10.21741/9781644903612-18

Figure 2 Experimental setup for performing tensile test

The three-level control factors selected for the experimental analysis are detailed in Table 3. A Taguchi L9 orthogonal array was utilized to examine the effects of various factors on the mechanical properties of honeycomb and cubic infill patterns. The experimental design, based on Taguchi's method, was developed and analyzed using MINITAB 19. The optimal combination of control factors was determined by converting the experimental results into signal-to-noise (S/N) ratios. The larger is better feature can be used to indicate the S/N ratio for maximal Young's modulus and ultimate tensile strength. The Signal to Noise (S/N) ratio greater is better characteristic equation is displayed beneath Equation 1.

$$\frac{S}{N} = -10 Log \left(\frac{\sum \frac{1}{Y^2}}{n} \right) \qquad Eq.\,1\,[16]$$

Where n is the number of observations and y is the observed data

Table 3. Control factors and levels used in the experiment.

Factors	Level 1	Level 2	Level 3
Layer height (mm)	0.15	0.2	0.3
Lay-up speed (mm/min)	40	60	80
Infill density (%)	20	40	60

Result & Discussion
Mechanical Properties
The process variables, such as layer height, layup speed, and infill density, affect the mechanical characteristics of the 3D printed items. The findings in Table 4 make it abundantly evident that 3D items printed at lower speeds and heights and with a higher infill density have better mechanical qualities than those printed at higher speeds and heights.

Table 4. Representation of Tensile properties and control factors

Sample	Layer height (mm)	Lay-up speed (mm/Min)	Infill Density (%)	Honeycomb		Cubic	
				Young's Modulus (MPA)	Ultimate Tensile strength (MPA)	Young's Modulus (MPA)	Ultimate Tensile strength (MPA)
1	0.15	40	20	878	32.4	974	32.9
2	0.15	60	40	942	33.4	977	37.1
3	0.15	80	60	1176	35.6	1150	42.6
4	0.2	40	20	891	32.8	871	32.6
5	0.2	60	40	947	33.2	924	35.7
6	0.2	80	60	983	35.5	997	34.5
7	0.3	40	20	902	35.4	854	31.6
8	0.3	60	40	998	36.5	912	32.2
9	0.3	80	60	945	34.8	974	33.7

There is a strong correlation between process parameters and mechanical performance, according to the results in Table 4. Different performance trends influenced by layer height, lay-up speed, and infill density are revealed when comparing the Young's modulus of cubic and honeycomb infill patterns. Honeycomb infill has a slightly higher Young's modulus than cubic at lower infill densities (20%), indicating its greater stiffness-to-weight efficiency with less material consumption. Sample 1, for example, displays 878 MPa for honeycomb and 974 MPa for cubic. Nevertheless, cubic infill reaches competitive or superior stiffness when the infill density rises to 40% and 60%. Figure 3 shows a peak Young's modulus of 1150 MPa as opposed to 1176 MPa for honeycomb (Sample 3). These patterns show that cubic infill works better in situations demanding more structural strength at increasing densities, even though honeycomb is initially beneficial for lightweight designs.

Process variables such as layer height and lay-up speed are critical in determining the Young's modulus for both honeycomb and cubic infill patterns. A higher lay-up speed (80 mm/min) and a thinner layer height (0.15 mm) for cubic infill help to improve material consolidation, which raises stiffness. However, honeycomb infill performs quite consistently over a range of parameters, indicating that it is appropriate for applications with varying production conditions. Overall, honeycomb works well in lightweight designs that prioritize stiffness efficiency, whereas cubic infill provides a fair trade-off between strength and stiffness, making it perfect for applications needing both qualities.

Materials Joining and Manufacturing Processes: MJMP 2025
Materials Research Proceedings 55 (2025) 118-129

Materials Research Forum LLC
https://doi.org/10.21741/9781644903612-18

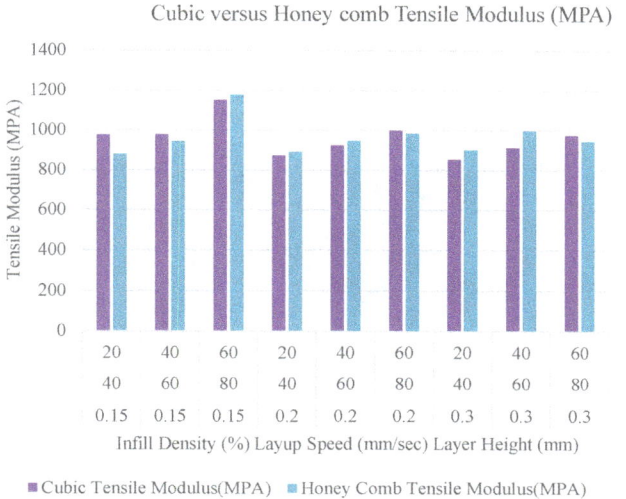

Figure 3 Cubic versus Honey comb Tensile Modulus (MPA)

Cubic infill continuously performs better than honeycomb in terms of ultimate tensile strength at different layer heights, lay-up speeds, and infill densities. With a maximum tensile strength of 42.6 MPa (Sample 3) compared to 36.5 MPa for honeycomb (Sample 8) in Figure 4, the superiority of cubic infill becomes even more noticeable at increasing infill densities. This pattern shows that cubic infill is the best option for load-bearing applications since it can sustain more loads before failing. Cubic infill's slight tensile strength advantage over honeycomb, even at lower densities (20%), emphasizes its structural integrity independent of material utilization.

The robustness of cubic infill is further highlighted by the impact of production factors. By guaranteeing tighter layer adhesion and less void formation, a lower layer height (0.15 mm) and a faster lay-up speed (60–80 mm/min) increase the tensile strength. Conversely, honeycomb infill exhibits modest gains in tensile strength as infill density rises, but it falls short of cubic infill in all circumstances. According to these results, cubic infill is the best alternative for situations where tensile performance is a top priority, while honeycomb is still a good choice for designs that prioritize material economy and moderate strength.

Materials Joining and Manufacturing Processes: MJMP 2025   Materials Research Forum LLC
Materials Research Proceedings 55 (2025) 118-129   https://doi.org/10.21741/9781644903612-18

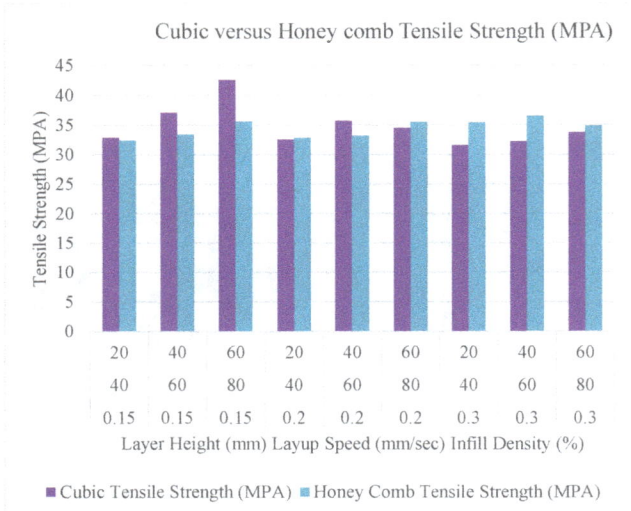

Figure 4 Cubic versus Honey comb Tensile Strength

The data show that PLA parts printed at lower layer heights and slower lay-up speeds consistently displayed superior mechanical properties compared to those printed at higher settings[17]. This improvement can be attributed to increased curing and settling time, which enhances layer bonding and structural integrity.

Analysis of Mechanical properties using Taguchi design of experiments
Tables 5 and 6 show the ultimate tensile strength and Young's modulus for the honeycomb and cubic infill patterns, as well as the related S/N ratios. The S/N ratio graphs for the several control factors influencing the ultimate tensile strength and Young's modulus of the honeycomb and cubic infill patterns are shown in Figures 5 and 6, respectively.

Table 5. Experimental design parameters for Honey Comb

Honeycomb							
Exp. No.	Layer height	Lay-up speed	Infill density	Young's modulus	S/N ratio	Ultimate tensile strength	S/N ratio
1	0.15	40	20	878	58.8699	32.4	30.2109
2	0.15	60	40	942	59.4810	33.4	30.4749
3	0.15	80	60	1176	61.4081	35.6	31.0290
4	0.20	40	40	891	58.9976	32.8	30.3175
5	0.20	60	60	947	59.5270	33.2	30.4228
6	0.20	80	20	983	59.8511	35.5	31.0046
7	0.30	40	60	902	59.1041	35.4	30.9801
8	0.30	60	20	998	59.9826	36.5	31.2459
9	0.30	80	40	945	59.5086	34.8	30.8316

Materials Joining and Manufacturing Processes: MJMP 2025　　　　Materials Research Forum LLC
Materials Research Proceedings 55 (2025) 118-129　　　　https://doi.org/10.21741/9781644903612-18

Table 6. Experimental design parameters for Cubic

Cubic							
Exp. No.	Layer height	Lay-up speed	Infill density	Young's modulus	S/N ratio	Ultimate tensile strength	S/N ratio
1	0.15	40	20	974	59.77	32.9	30.34
2	0.15	60	40	977	59.80	37.1	31.39
3	0.15	80	60	1150	61.21	42.6	32.59
4	0.20	40	40	871	58.80	32.6	30.26
5	0.20	60	60	924	59.31	35.7	31.05
6	0.20	80	20	997	59.97	34.5	30.76
7	0.30	40	60	854	58.63	31.6	29.99
8	0.30	60	20	912	59.20	32.2	30.16
9	0.30	80	40	974	59.77	33.7	30.55

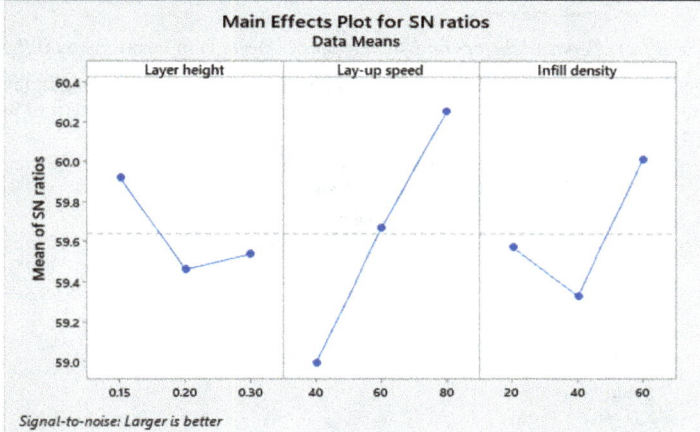

Figure 5. Effect of control factors on Young's modulus of honeycomb infill pattern

By analyzing these factors, one can effectively optimize the control parameters to achieve the maximum Young's modulus and ultimate tensile strength. This optimization is critical for tailoring the mechanical performance of printed components to specific application requirements. From Figure 5, it is evident that the combination of a 1.5 mm layer height, 80 mm/min lay-up speed, and 60% infill density produces the highest Young's modulus. This combination likely ensures optimal bonding between layers and efficient load distribution within the infill structure, enhancing stiffness. Additionally, the higher infill density provides better structural integrity and minimizes internal voids, further contributing to improved mechanical properties. This insight underscores the importance of precise parameter selection in achieving desired outcomes in 3D-printed parts.

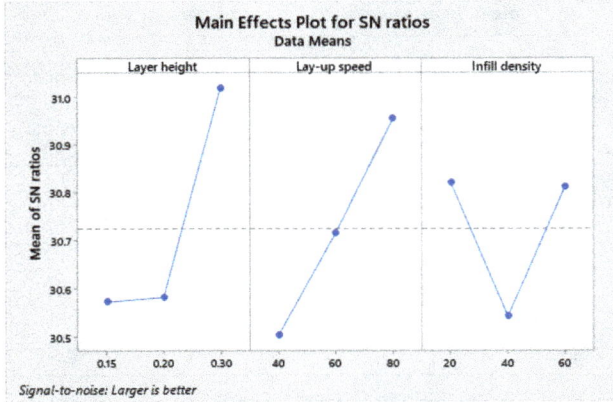

Figure 6. Effect of control factors on Ultimate tensile strength of honeycomb infill pattern

From the figure 6, it is clear that the factor combination of 0.3 mm layer height, lay-up speed of 80 mm/min and 20% of infill density yield maximum Ultimate tensile strength of honeycomb infill pattern.

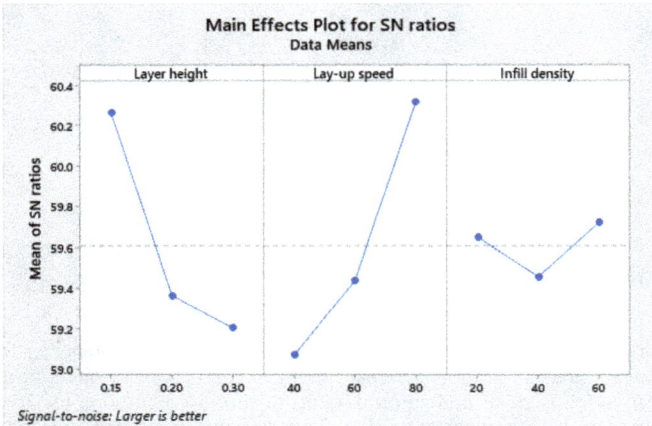

Figure 7. Effect of control factors on Young's modulus of cubic infill pattern

Materials Joining and Manufacturing Processes: MJMP 2025
Materials Research Proceedings 55 (2025) 118-129

Materials Research Forum LLC
https://doi.org/10.21741/9781644903612-18

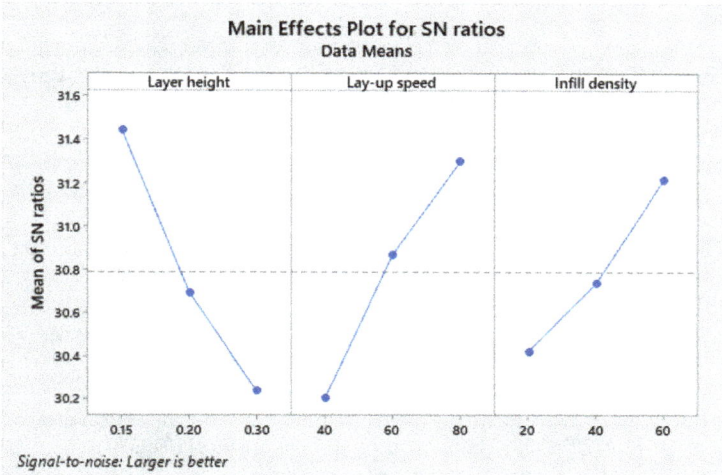

Figure 8. Effect of control factors on Ultimate tensile strength of cubic infill pattern

From Figures 7 and 8, it is evident that the control factor combination of a 0.15 mm layer height, 80 mm/min lay-up speed, and 60% infill density yields the maximum Young's modulus and ultimate tensile strength for the cubic infill pattern. This combination likely provides a fine balance between precision and structural integrity. The smaller layer height ensures better surface finish and stronger interlayer adhesion, which directly enhances tensile properties. The higher lay-up speed optimizes the deposition process, maintaining uniformity without compromising the bonding quality. Additionally, the 60% infill density provides sufficient internal support and minimizes voids, contributing to the superior mechanical performance. This combination highlights the significance of carefully chosen parameters for achieving optimal mechanical properties in cubic infill designs, making it suitable for applications requiring high strength and durability.

Conclusions

The comparison of cubic and honeycomb infill patterns makes it clear how process variables affect mechanical performance in 3D-printed PLA products. With a Young's modulus of 878 MPa as opposed to 974 MPa for cubic at 20% density, honeycomb infill performs exceptionally well in terms of stiffness-to-weight efficiency at lower infill densities. Nevertheless, cubic infill outperforms honeycomb at peak values of 1150 MPa and 1176 MPa, respectively, when infill density rises to 40% and 60%.

In terms of ultimate tensile strength, cubic infill routinely performs better than honeycomb at all density levels. Cubic achieves a tensile strength of 42.6 MPa at maximum infill density (60%), whereas honeycomb achieves 36.5 MPa.

Higher lay-up speeds (80 mm/min) and smaller layer heights (0.15 mm) enhance material consolidation, increasing stiffness and tensile strength. The 0.15 mm layer height, 80 mm/min lay-up speed, and 60% infill density for cubic infill were shown to be the best control factor combinations for optimizing both tensile strength and Young's modulus.

Materials Joining and Manufacturing Processes: MJMP 2025
Materials Research Proceedings 55 (2025) 118-129

Materials Research Forum LLC
https://doi.org/10.21741/9781644903612-18

References

[1] A. Y. Alqutaibi, M. A. Alghauli, M. H. A. Aljohani, M. S. Zafar, Advanced additive manufacturing in implant dentistry: 3D printing technologies, printable materials, current applications and future requirements, Bioprint (2024) e00356. https://doi.org/10.1016/j.bprint.2024.e00356

[2] A. Jandyal, I. Chaturvedi, I. Wazir, A. Raina, M. I. U. Haq, 3D printing–A review of processes, materials and applications in industry 4.0, Sust. Oper. Comput. (2022) 33-42. https://doi.org/10.1016/j.susoc.2021.09.004

[3] K. B. Mustapha, K. M. Metwalli, A review of fused deposition modelling for 3D printing of smart polymeric materials and composites, Eur. Polym. J. 156 (2021) 110591. https://doi.org/10.1016/j.eurpolymj.2021.110591

[4] R. H. Risad, M. H. Ahmed, A. Basher, S. Rashid, M.M.A. Shishir, K.R. Hossain, FDM printing process and its Biomedical Application.Chem. Res. Technol. 1.3 (2024) 138-149. https://doi.org/10.22034/chemrestec.2024.467346.1021

[5] E. H. Tümer, H. Y. Erbil, Extrusion-based 3D printing applications of PLA composites: a review, Coatings 11.4 (2021) 390. https://doi.org/10.3390/coatings11040390

[6] V. Cojocaru, D. Frunzaverde, C. O. Miclosina, G. Marginean, The influence of the process parameters on the mechanical properties of PLA specimens produced by fused filament fabrication A review, Polymers 14.5 (2022) 886. https://doi.org/10.3390/polym14050886

[7] H. Radhwan, Z. Shayfull, S. M. Nasir, A.R. Irfan, Optimization parameter effects on the quality surface finish of 3D-printing process using taguchi method, IOP Conf. Ser.: Mater. Sci. Eng. Vol. 864. No. 1. (2020) DOI 10.1088/1757-899X/864/1/012143

[8] M. Q. Tanveer, G. Mishra, S. Mishra, R. Sharma, Effect of infill pattern and infill density on mechanical behaviour of FDM 3D printed Parts-a current review, Mater. Today: Proc. 62 (2022) 100-108. https://doi.org/10.1016/j.matpr.2022.02.310

[9] K. P. Aveen, F. Vishwanath Bhajathari, C. J. Sudhakar, 3D Printing & Mechanical Characteristion of polylactic acid and bronze filled polylactic acid components, IOP Conf. Ser. Mater. Sci. Eng. Vol. 376. No. 1. IOP Publishing (2018) DOI 10.1088/1757-899X/376/1/012042

[10] B. Aloyaydi, S. Sivasankaran, A. Mustafa, Investigation of infill-patterns on mechanical response of 3D printed poly-lactic-acid. Polym. Test. 87 (2020) 106557. https://doi.org/10.1016/j.polymertesting.2020.106557

[11] L. Dezaki, Mohammadreza, Mohd Khairol Anuar Mohd Ariffin. The effects of combined infill patterns on mechanical properties in fdm process, Polymers 12.12 (2020) 2792. https://doi.org/10.1016/j.matpr.2022.09.227

[12] P. Sanket, N. Zaveri, Investigating the influence of infill pattern on the compressive strength of fused deposition modelled PLA parts, Proceedings of: ICIMA 2020. Springer Singapore, (2020) https://link.springer.com/chapter/10.1007/978-981-15-4485-9_25

[13] R. Srinivasan, W. Ruban, A. Deepanraj, R. Bhuvanesh, T. Bhuvanesh, Effect on infill density on mechanical properties of PETG part fabricated by fused deposition modelling, Mater. Today: Proc.27 (2020) 1838-1842. https://doi.org/10.1016/j.matpr.2020.03.797

[14] K. Sharma, K. Kumar, K. R. Singh, M. S. Rawat, Optimization of FDM 3D printing process parameters using Taguchi technique, IOP Conf. Ser.: Mater. Sci. Eng. Vol. 1168. No. 1. IOP Publishing, 2021. DOI 10.1088/1757-899X/1168/1/012022

[15] K. P. Aveen, V. Bhajantri, R. D'Souza, N. V. Londe, S. Jambagi, Experimental analysis on effect of various fillers on mechanical properties of glass fiber reinforced polymer composites, AIP Conf. Proc. Vol. 2057. No. 1. (2019). https://doi.org/10.1063/1.5085615

[16] Tseng, Wan-Tsun, Chen-Nan Kuo, Li-Iau Su, Optimizing design parameters of a novel PM transverse flux linear motor, Trans. Can. Soc. Mech. Eng. 39.3 (2015) 443-454. https://doi.org/10.1139/tcsme-2015-0033

[17] U. V. Akhil, N. Radhika, B. Saleh, S. Aravind Krishna, N. Noble, L. Rajeshkumar, A comprehensive review on plant-based natural fiber reinforced polymer composites: fabrication, properties, and applications, Polym. Compos. 44.5 (2023) 2598-2633. https://doi/abs/10.1139/tcsme-2015-0033

Materials Joining and Manufacturing Processes: MJMP 2025   Materials Research Forum LLC
Materials Research Proceedings 55 (2025) 130-135   https://doi.org/10.21741/9781644903612-19

Joining of aluminum tube to PVC pipes through electromagnetic force

Shanthala K.[1,a] *, Gajanan M. Naik[1,b], Akshit Kochhar[1,c],
Ramesh S.[2,d], Sadashiv Bellubbi[3,e]

[1]Department of Mechanical Engineering, RV Institute of Technology and Management, JP-Nagar, Bengaluru 560076, Affiliated to Visvesvaraya Technological University, Belagavi. Karnataka, India

[2]School of Computer Science and Engineering, RV University, Mysore Rd, RV Vidyaniketan Post Bengaluru 560058, Karnataka, India

[3]Department of Mechanical Engineering, Jain College of Engineering and Research, Udyambag-590008, Belagavi. Affiliated to Visvesvaraya Technological University, Belagavi. Karnataka, India

[a]kshanthala2@gmail.com, [b]gajamnaik@gmail.com, [c]rvit21bme027.rvitm@rvei.edu.in, [d]ramnitk2016@gmail.com, [e]bellubbisadashiv@gmail.com

Keywords: Electromagnetic, Joining, Aluminium, PVC, Residual Stress

Abstract. This study investigates the joining of Al to polyvinyl chloride (PVC) pipes using electromagnetic force, focusing on the role of discharge energy and residual stress in optimizing the bond quality. The method addresses challenges faced in traditional joining techniques, offering a promising alternative by using electromagnetic fields to induce localized heating and pressure. Results demonstrated that medium and high discharge energies, ranging from 13.5 kV to 14.5 kV, produced smooth, gap-free interfaces with enhanced mechanical bonding. Existence of residual compressive stresses at the interface contributed to improved bond strength and durability. The findings emphasize the importance of discharge energy and residual stress in achieving high-quality metal-to-polymer joints, highlighting key parameters for optimizing electromagnetic welding processes.

1. Introduction

The joining of dissimilar materials, such as metals and polymers, is a critical challenge in manufacturing industries due to vast differences in their mechanical and thermo-physical properties [1]. Existing welding methods are generally ineffective for such applications, particularly for metal-polymer joining, due to these material disparities.

Fusion welding of polymers demands precise control of parameters, limiting its feasibility for large-scale production. In response to these limitations, solid-state welding techniques, such as Electromagnetic Welding (EMW), have emerged as a viable alternative for joining dissimilar materials [2].

EMW is an advanced solid-state welding technique that uses high-energy electromagnetic fields to join materials. It offers several advantages, involves the ability to join metals to non-metals without melting the materials, thus avoiding thermal degradation. This technique is particularly suitable for lightweight structural applications and challenging combinations, such as metal-to-polymer or metal-to-composite joints [3]. The present study focuses on the use of EMW to produce joints between Al and polymer-based materials, emphasizing the influence of welding factors on joint quality, mechanical strength, and residual stresses.

EMW has been thoroughly studied while joining various materials, particularly metals that are difficult to weld using conventional methods. For example, it has been certainly applied to join dissimilar metals such as aluminium, copper, and nickel, overcoming the challenges posed by their

Content from this work may be used under the terms of the Creative Commons Attribution 3.0 license. Any further distribution of this work must maintain attribution to the author(s) and the title of the work, journal citation and DOI. Published under license by Materials Research Forum LLC.

Materials Joining and Manufacturing Processes: MJMP 2025   Materials Research Forum LLC
Materials Research Proceedings 55 (2025) 130-135   https://doi.org/10.21741/9781644903612-19

differing thermo-physical properties [4]. However, when metals are joined to polymer-based composites, additional challenges arise. Fusion-based welding techniques are often unsuitable due to the need for precise parameter control, which limits scalability [5]. Solid-state techniques, including EMW, provide a promising alternative for such applications [6-8]. EMW allows control over key parameters such as discharge energy, standoff distance, and impact angle, which are necessary to maintain the strength of joints.

Watanabe et al. [6] explored EMW for aluminium-to-metallic glass joints and demonstrated the vital needs for accurate control over impact force. Excessive force may lead to defects or damage, particularly in brittle materials like metallic glass. To address this, the usage of shock-absorbing plates made of annealed aluminium was proposed, which helped manage the impact force and prevent damage. Similarly, Aizawa et al. [7] investigated EMW for joining flexible-printed-circuit-boards (FPCB) using polyamide flyers. Due to flyers' low surface conductivity, a driver plate made of aluminium was introduced to facilitate the joining process. The results discovered that the characteristic wavy interface of EMW was observed only when the driver plate was used, highlighting the importance of managing welding parameters for achieving optimal joint quality.

The production of wavy interfaces, a hallmark of impact welding, is essential for ensuring mechanical interlocking and joint strength [8-9]. However, improper control of parameters like discharge energy and standoff distance can lead to defects like incomplete joints, wrinkling, or damage to the target material [10-11].

Despite extensive research on EMW for metal-to-metal and metal-to-composite joints, there is limited understanding of joining aluminium to polymer-based materials like PVC. Existing studies have focused on metallic glasses and FPCBs, with insufficient exploration of key parameters such as discharge energy, standoff distance, and impact force for metal-to-polymer combinations. Additionally, the role of residual stresses in joint integrity has been largely overlooked.

Although EMW has been used to various materials, the novelty of the present study lies in addressing the difficulties faced with joining aluminium to PVC combination that remains underexplored. This work aims to examine the effects of EMW parameters on aluminium-to-PVC joint quality, thus filling the identified research gap. By focusing on discharge energy, residual stresses, and mechanical interlocking, this research seeks to provide visions that can guide the optimization of EMW for metal-to-polymer joints.

2. Methodology
The experiment setup for the EMW of Al-to-PVC pipe joints utilized EMW machine rated at 5kJ (Make: Zeonics Bangalore), with a voltage range of 12.5kV to 14.5kV, adjustable through a control unit comprising a trigger unit, discharge unit, compressor, and a varying voltage system. The voltage range was chosen on the basis of preliminary trials, which provides sufficient energy to achieve joint formation without causing material damage or deformation beyond acceptable limits. The machine included three 15.5 μF capacitors connected in parallel, providing a total capacitance of 43.5 μF, with a peak current of 60 kA. During the EMW process, commercially available aluminium tube (Al-LM6) was lapped onto a PVC pipe, SOD maintained between the two materials was 0.5mm. The impact forces produced by the discharge energy were varied across three levels, with each discharge energy corresponding to specific voltage settings ranged from 12.5 kV to14.5 kV. Controlling factors for Al-LM6 / PVC pipe EMW are listed in Table 1.

After the joining process, the joints were undergone with mechanical strength testing, including tensile and shear tests, to assess their load-bearing capacity. Microscopic analysis was performed to examine the bonding between the Al-LM6/PVC, focusing on surface deformation, residual stress, and the frictional coefficient at the contact zone. Microstructural studies were done with an optical microscope (Nikon Eclipse LV150). Additionally, an oscilloscope (Tektronix TCPA400 series) was used to measure and analyse the discharge current and voltage waveforms, which were correlated with the joint quality to understand the effect of the discharge energy on the final weld.

Table 1. Controlling factors for Al-LM6 / PVC pipe joining

Sl. No.	Voltage [kV]	Current [kA]	Energy [kJ]
1.	12.5	51	3.3984
2.	13.5	55.5	3.9639
3.	14.5	60	4.5720

Samples were subjected to push-type shear tests using a Universal Testing Machine (Model UTE-60) with a 400 kN capacity to examine the strengths of the joints. Displacement-load curves were obtained to analyse the force required to separate the aluminium flyer from the PVC pipe. Residual stress findings in the joint were conducted using X-ray diffraction-based technique in a stress measurement system iXRD MGR40P, (Model: PROTO, CANADA). The d-spacing variations were analysed to determine strain, and standard XRD residual stress computation techniques were used to compute the residual stress.

3. Results and Discussion

The Al flyer and PVC targets are assembled in Fig. 1 prior to joining, and samples are joined at various discharge voltages shown in Table 1. The produced Al-LM6 tube's external appearance shows an identical profile across all three energy levels. When the flyer tube's striation was closely examined, it revealed a wrinkled surface at 12.5kV. At 13.5 kV, the flyer tube profile shown a smooth surface. There were clearer ridges and fewer wrinkles in the sample at 14.5kV. However, the increased impact force caused the PVC core to bend. By examining the cut portions, the joint's quality was further assessed (Fig. 2). Incomplete joints were seen in cut parts of the samples at lesser energy. There was a noticeable space between the jobs and a deficiency of frictional grip between the pipe and the Al tubing. It was observed that the jobs linked with different energy levels (medium and higher) had a smooth joint without gap at the Al-LM6/PVC pipe interface. The existence of ridges with a gap for jobs bonded at lesser energy was confirmed by OM images (Fig. 3).

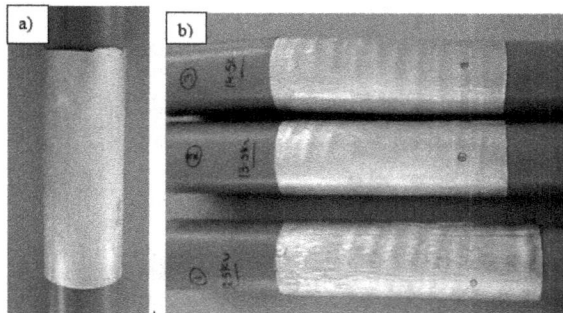

Fig. 1 Images of a) Flyer / Target Assembly b) Joined Samples

Materials Joining and Manufacturing Processes: MJMP 2025　　　　Materials Research Forum LLC
Materials Research Proceedings 55 (2025) 130-135　　　　https://doi.org/10.21741/9781644903612-19

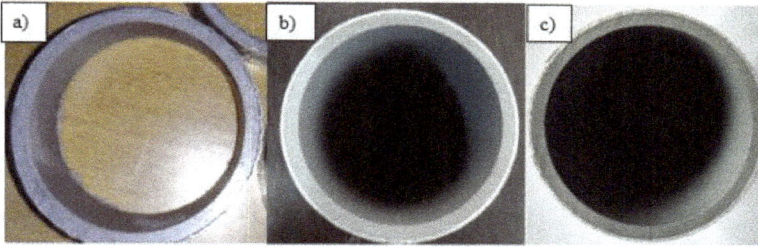

Fig. 2 Images of Al-LM6 / PVC pipe cut portions
a) 12.5kV b) 13.5kV c) 14.5kV

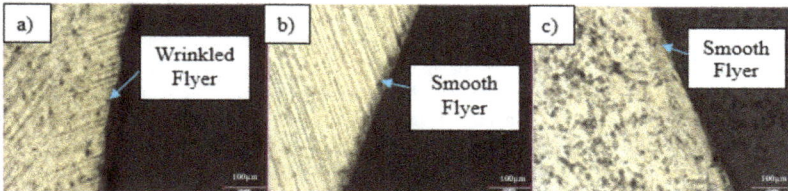

Fig. 3 Optical Microscope pictures of Al-LM6 / PVC pipe cut portions
a) 12.5kV b) 13.5kV c) 14.5kV

3.1 Strength assessment of the Al-LM6 / PVC joint by simple push test

A shear push examination in UTM was conducted to assess the strengthness of the Al-to-PVC pipe joint at a crosshead speed of 1 mm/min. Fig. 4 displays the load-displacement plot for the joint created at 13.5 kV, demonstrating that the lap joint could withstand a maximum shear load of 0.84 kN.

The curve exhibited irregular waviness, with the load increasing initially, dropping after some displacement and rising again before the joint sheared off at a lesser load. This variation is caused to ridges and crests observed on the aluminium flyer, which reduced the contacting area as displacement increased, causing the load to drop at the point of separation.

The findings underscore the importance of discharge energy in determining joint quality and strength in electromagnetic welding. At 13.5 kV, the mechanical interlocking between aluminium and PVC, enhanced by compressive forces during welding, contributed to joint integrity. However, the waviness in the load-displacement curve suggests localized variations in impact forces and friction, potentially affecting joint uniformity.

Materials Joining and Manufacturing Processes: MJMP 2025   Materials Research Forum LLC
Materials Research Proceedings 55 (2025) 130-135   https://doi.org/10.21741/9781644903612-19

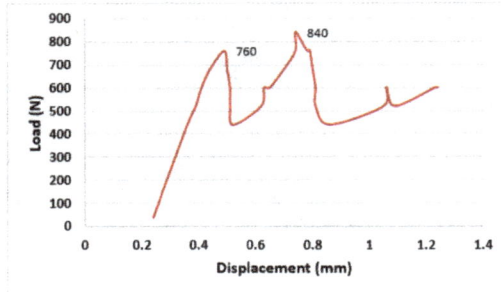

Fig. 4 Load-Displacement curve

3.2 Residual stress

Al-LM6 and PVC pipes are not appropriate for metallurgical joints. Rather, the frictional coefficient—the amount of residual stress in the contact zone—determines the strongness of the Al/PVC lapped interface. 1.3 MPa tensile residual stress was used to compute the parental aluminium. With an average compressive residual stress value of -12.6MPa, -19.7MPa and -25MPa at 12.5 kV, 13.5 kV and 14.5 kV respectively, the lapped area under the electromagnetic compressive force was observed.

The residual stress measurements validate the relationship between the discharge energy and the mechanical properties of the joint. The increased compressive stresses at higher energy levels suggest that greater impact forces lead to stronger mechanical interlocking at the aluminium-to-PVC interface. This is consistent with the observed increase in joint strength, as evidenced by the shear test results. Moreover, the comparison with similar studies on aluminium-to-polyurethane joints reinforces the effectiveness of electromagnetic welding for joining aluminium to rigid polymers like PVC, which cannot be welded using conventional methods.

4. Conclusion

Joining aluminium to PVC pipe through electromagnetic force was demonstrated successfully in this study, and the following conclusions are drawn:

- Discharge energy plays a critical role in achieving high-quality joints, with discharge energy at 13.5 kV and 14.5kV produced smooth, gap-free interfaces with stronger mechanical bonds.
- Higher discharge energies resulted in increased compressive residual stresses (e.g., -12.6MPa, -19.7MPa and -25MPa at 12.5kV, 13.5kV and 14.5kV respectively), enhancing mechanical interlocking between aluminium and PVC.
- The increased compressive stresses suggest that higher discharge energies lead to stronger bonds, emphasizing the importance of discharge energy in optimizing the electromagnetic welding parameters for metal-to-polymer joints.

The study provides a foundation for optimizing electromagnetic welding parameters for metal-to-polymer joints. Further investigation into the physics of electromagnetic welding for various material combinations is made possible by the insights gained from this work.

References

[1] S. Razzaq, Z.X. Pan, H.J. Li, S.P. Ringer, and X.Z. Liao, Joining dissimilar metals by additive manufacturing: A review, J. Mater. Res. Technol. 31 (2024) 2820-2845. https://doi.org/10.1016/j.jmrt.2024.07.033

[2] K. Shanthala, T. N. Sreenivasa, H. N. N. Murthy, S. Dond, H. Choudhury, and A. Sharma, Joining of tubular steel-steel by unconventional magnetic pulse force: environmentally friendly technology, Bull Mater. Sci 42 (2019) 260. https://doi.org/10.1007/s12034-019-1947-5

[3] D. Pereira, J. P. Oliveira, T. G. Santos, R. M. Miranda, F. Lourenço, J. Gumpinger, and R. Bellarosa, Aluminium to Carbon Fibre Reinforced Polymer tubes joints produced by magnetic pulse welding, Compos. Struct. 230 (2019) 111512. https://doi.org/10.1016/j.compstruct.2019.111512

[4] K. Shanthala, and T. N. Sreenivasa, Review on electromagnetic welding of dissimilar materials, Front. Mech. Eng. 11 (2016) 363-373. https://doi.org/10.1007/s11465-016-0375-0

[5] A. Yousefpour, M. Hojjati, and J-P. Immarigeon, Fusion bonding/welding of thermoplastic composites, J. Thermoplast. Compos. Mater. 17(4) (2004) 303-341. https://doi.org/10.1177/0892705704045187

[6] M. Watanabe, S. Kumai, G. Hagimoto, Q. Zhang, and K. Nakayama, Interfacial microstructure of aluminum/metallic glass lap joints fabricated by magnetic pulse welding, Mater. Trans. 50 (2009) 1279-1285. https://doi.org/10.2320/matertrans.ME200835

[7] T. Aizawa, K. Okagawa, and M. Kashani, Application of magnetic pulse welding technique for flexible printed circuit boards (FPCB) lap joints, J. Mater. Proc. Technol. 213 (2013) 1095-102. https://doi.org/10.1016/j.jmatprotec.2012.12.004

[8] W.S. Hwang, N.H. Kim, H.S. Sohn, and J.S. Lee, Electromagnetic joining of aluminium tubes on polyurethane cores, J. Mater. Proc. Techno. 34,1-4 (1992) 341-348. https://doi.org/10.1016/0924-0136(92)90126-D

[9] S. Patra, K. S. Arora, M. Shome, and S. Bysakh, Interface characteristics and performance of magnetic pulse welded copper-Steel tubes, J. Mater. Proc. Techno. 245 (2017) 278-286. https://doi.org/10.1016/j.jmatprotec.2017.03.001

[10] S. Bellubbi, and N. Sathisha, Reduction in through put time of drum shell manufacturing by single-V welding configuration, IOP Conf. Series: Mater. Sci. Eng. 376 (2018) 012104. https://doi.org/10.1088/1757-899X/376/1/012104

[11] M. Hahn, C. Weddeling, J. Lueg-Althoff, and A.E.Tekkaya, Analytical approach for magnetic pulse welding of sheet connections, J. Mater. Proc. Technol. 230 (2016) 131-142. https://doi.org/10.1016/j.jmatprotec.2015.11.021

Materials Joining and Manufacturing Processes: MJMP 2025
Materials Research Proceedings 55 (2025) 136-140

Materials Research Forum LLC
https://doi.org/10.21741/9781644903612-20

Lead-free solders for high-temperature applications

SATYANARAYAN[1, a *], K. NARAYAN PRABHU[2,b]

[1]Department of Mechanical Engineering, Alva's Institute of Engineering and Technology Karnataka, Moodubidire, Mangalore 574225, India (Affiliated to VTU Belgaum)

[2]Department of Metallurgical and Materials Engineering, National Institute of Technology Karnataka, Surathkal, Mangalore 575025, India

[a]satyan.nitk@gmail.com, [b]knprabhu@nitk.edu.in

Keywords: Pb-Free Solders, Alloying, Wetting, High Temperature

Abstract. Due to the toxicity of Pb, efforts to develop alternatives to Pb-based solders have been increased significantly. However, only a limited number of lead (Pb)-free solder systems exist for high-temperature applications. In the present paper, a review of Pb-free solder alloys for high-temperature (300°C–400°C) applications is carried out. Recent research and developments in the field of high-temperature solder alloys is highlighted in the current review. The solder systems with alloying additions like Au, Sb, Ni and Zn are some of the potential candidate materials for high-temperature applications in place of Pb-based solders.

Introduction

The soldering technique in flip-chip connections and ball-grid arrays (BGA) is vital in electronic applications. Solder joints in the printed circuit boards (PCBs) serve as mechanical and electrical connections. The solder joints are thought to fail when electrical or mechanical connections are out of service. At present, the soldering technologies widely used for the mass production of circuits are plated-through-hole (PTH) and surface mount technology (SMT) [1].

The lead (Pb) containing solders [mainly eutectic Pb-Sn] are widely used in PCBs/electronic applications. However, Pb in the solder material is highly toxic and considered hazardous to the environment [2]. European Commission directives, i.e., WEEE (Waste Electrical and Electronic Equipment), RoHS (Restriction of Use of Hazardous Substances), and ELV (End of Life Vehicle) directives, have restricted the use of Pb in electronic applications. Even Japan passed the 'Home Electronics Recycle Law' in 1998 [3]. These initiatives have led to the development of novel lead-free solders such as Sn–Bi, Sn–In, Sn–Zn, Sn–Cu, Sn–Ag, Sn–Ag–Zn, Sn–Zn–Bi, Sn–Ag–Cu, Sn–Ag–Bi for electronic applications in which Sn is a significant element [4-9]. Legislative initiatives have asked to redesign electronic products using Pb-free components.

It is reported that high lead-bearing solders are still in use even though Integrated Circuit (IC) boards are assembled with middle-temperature range Pb-free solder alloys i.e., Sn-Ag-Cu alloy [10]. Hence, establishing high-temperature lead-free solders is an urgent priority in the electronics industry.

Only a few high-temperature Pb-free solders are increasingly being used for high-performance systems. However, few papers and reports on the research and development for high-temperature lead-free solders are available [9, 11-13]. In the present review paper, an attempt has been made to understand the recent development in high-temperature lead-free soldering for electronic packaging applications.

Content from this work may be used under the terms of the Creative Commons Attribution 3.0 license. Any further distribution of this work must maintain attribution to the author(s) and the title of the work, journal citation and DOI. Published under license by Materials Research Forum LLC.

Materials Joining and Manufacturing Processes: MJMP 2025
Materials Research Proceedings 55 (2025) 136-140

Materials Research Forum LLC
https://doi.org/10.21741/9781644903612-20

High temeperature Pb-free solders

High-temperature solders are used as die-attach solders for assembling optical components, automobile circuit boards, circuit modules, etc [11]. For selecting Pb free-solders for high-temperature applications, the essential requirements are as follows [12-13]

- Melting temperature should be between 260°C to 400°C.
- Minor volume expansion
- Should be drawn as thin wires or sheets
- Good electrical and thermal conductivities
- Good mechanical properties (especially resistance to the fatigue failure)
- Flux less

Among these, significant properties for selecting high-temperature solders are Melting temperature, the coefficient of thermal expansion CTE, and Modulus. Table 1 provides information on the properties of high-temperature Pb-based solders. A few selected properties of high-temperature Pb-solders are mentioned in Table 2.

Table 1: Few properties of selected high-temperature solders [10].

Major Solders	Applications	Major Requirements
High Pb alloys	Flip-chip packages	Reflow resistance, Low-ray emission
	Internal connection of passive components	Reflow resistance
	Quartz device	Reflow resistance, Airtightness
	Heat-resistant vehicle packages	Thermal fatigue resistance Heat exposure resistance
	Module packages step soldering	Reflow resistance
High Pb alloys Au alloys	Die –attachment of power semiconductors and Application-specific integrated circuit (AISC)	Reflow resistance High thermal/electric conductivity Thermal fatigue resistance

Table 2: Current high-temperature solders and their properties [10].

Solder composition	Melting temperatue (°C)	CTE (ppm/°C)	Modulus (GPa)
Au20Sn	280	16	59
Au12Ge	356	13	72
Au3Si	363	12	83
Bi11Ag0.05Ge	262–360	–	–
Ag20In	695	–	–
Zn–(4–6)Al(–Ga, Ge, Mg, Cu)	~380	–	–
Zn–(10 to 30)Sn	360		

The melting behavior is also essential when designing new lead-free high-temperature solder alloys. In this regard, it is necessary to consider the proper regime of melting temperature. It is reported that the solidus temperature of the high-temperature solder should be at least 50°C higher than the melting point of the solder used in the second-level packaging to withstand peak temperatures of the second-level soldering [14].

The melting point of the commonly used Pb-free solder for second-level packaging (Sn-Ag-Cu) is 217°C [15]. Thus, the preferred melting temperature of a potential high-temperature lead-free solder by the industries is 270°C. Solder alloy's liquidus temperature should be below 350°C.

To avoid thermal degradation of polymers commonly used in the substrate as dielectric materials. Hence, industries set the melting temperature range from 270°C to 350°C to ensure effective process control.

Binary or ternary element additions also increase the melting range and modify the specific properties of solder alloys. For example, Nickel (Ni) increases the alloy's strength by forming intermetallic/intermediate/secondary phases with the Sn and increases an alloy's melting temperature. This is due to the marginal solubility of Ni in Tin [16]. Ni additions also slightly increased the solidus temperature and melting temperature of SAC155 Pb-free solder alloy [17].

Antimony (Sb) is soluble in tin. Sb addition to Sn increases the alloy strength; however, it slightly raises the alloy's melting temperature. Sn-6.2 wt%Sb is a peritectic composition alloy and its melting temperature is 243 °C, about 11 °C higher than that of Sn, 232 °C [18]. Bismuth (Bi) is also soluble in Sn, but it increases its strength and decreases the melting temperature enough to compensate for the effects of the additions of Sb and Ni [16].

Gold (Au) based solders are also called as hard solders. These hard solder materials are widely accepted for use in high-temperature electronics. It is reported that, among Au-based solders, the Au–Sn-based solders are the most commonly used for high-temperature electronic applications, including microwave devices, laser diodes, radio-frequency (RF) power amplifiers, and flip-chip bonding applications [16]. However, Au–Sn is very stiff and transfers most of the thermo-mechanical stress to the die, making it viable only for small die applications. In its pure state, this solder is limited to use at temperatures less than 280 °C, i.e., T<280°C [16]. The high price of Au has initiated the development of cheaper and more abundantly available material (such as Zinc) conducive to a mass manufacturing scenario.

Due to the high costs associated with Au alloys, silver (Ag) offers a possible solution for high-temperature applications. Silver has the best electrical conductivity and the second-best thermal conductivity. It also demonstrates high-temperature stability and has good mechanical properties [19]. Sn-Ag-Cu (SAC) are cheap and are the best Pb-free solder alloys for high-temperature soldering applications. However, it is essential to note that, too much Ni and Cu additions to the most SnAgCu solder alloys, can affect the alloy's wetting behaviour and its fluidity. This consideration is critical in selective as well as wave soldering techniques, which typically gain Ni as well as Cu elements as the metals dissolve from the Printed Wire Board (PWB) and surfaces component during processing [16].

Zinc (Zn) based alloys are used as high-temperature die attach materials and have gained popularity due to their low cost. By alloying Al and Sn to Zinc, the melting temperature of Zinc (419.5°C) can be lowered to a suitable range of 300–400°C. The absence of IMCs in the Zn–Sn alloy system makes the Zn–Sn solder alloys extremely ductile [16, 20].

Summary

The quest to replace Pb-containing solders for high-temperature applications is ongoing. Therefore, it is important to design and develop innovative, cost-effective, Pb-free solder alloys with superior properties in all aspects for electronic packaging applications in the soldering industry. Although Sn-Ag-Cu alloys are extensively used in high-temperature soldering applications, further research should be directed towards developing cost effective Au/Ni/Sb/Ni based solder composites with high melting temperature, good wettability and compatibility with the substrates, high mechanical strength, good oxidation resistance, adequate ductility and toughness.

References

[1] P Williams, Surface Mount Council Status Of The Technology Industry Activities And Action Plan, August 1999 https://www.ipc.org/4.0_Knowledge/4.1_Standards/smcstatus.pdf (accessed on 10/12/2019).

[2] Satyanarayan, KN Prabhu,Reactive wetting, evolution of interfacial and bulk IMCs and their effect on mechanical properties of eutectic Sn-Cu solder alloy, Adv.Coll. Int. Sci. 166 (2011) 87-118. https://doi.org/10.1016/j.cis.2011.05.005

[3] H. Jennie, Implementing lead-free electronics,McG.Hill.(2005) 10-90.

[4] Seo, Sun-Kyoung, S. K. Kang, D. Y Shih, H. M. Lee, An investigation of microstructure and microhardness of Sn-Cu and Sn-Ag solders as functions of alloy composition and cooling rate, J.electro.mat. 38 (2009) 257-265. https://doi.org/10.1007/s11664-008-0545-x

[5] G Kumar, KN Prabhu, Review of non-reactive and reactive wetting of liquids on surfaces, Adv. Coll.int. sci. 133 (2) (2007) 61-89. https://doi.org/10.1016/j.cis.2007.04.009

[6] KN Prabhu, M Varun, Satyanarayan, Effect of purging gas on wetting behavior of Sn-3.5Ag lead-free solder on nickel coated aluminium substrate, J.Mat.Eng.Per. 22 (2013) 723-728. https://doi.org/10.1007/s11665-012-0339-4

[7] R Mayappan, A.B Ismail, Z A Ahmed, T Ariga, LB Hussain, Wetting properties of Sn-Pb,Sn-Zn and Sn-Zn-Bi lead free solders, J.Tek. 46 (2007) 1-14.

[8] Y Y Chen, J G Duh, B S Chiou, The effect of substrate surface roughness on the wettability of Sn-Bi solders, J.mat. sci: mat. Ele.11 (2000) 279-83. https://doi.org/10.1023/A:1008917530144

[9] K Ales, D Andersson, N Hoo, J Pearce, A Watson, A Dinsdale, Stuart Mucklejohn,Current problems and possible solutions in high-temperature lead-free soldering, J.Mat.Eng.Per.21(5) (2012)629-637. https://doi.org/10.1007/s11665-012-0125-3

[10] M Sandeep, E George, M Osterman, M Pecht, High lead solder (over 85%) solder in the electronics industry: RoHS exemptions and alternatives J.mat. sci: mat. Ele.26 (6) (2015) 4021-4030. https://doi.org/10.1007/s10854-015-2940-4

[11] M Prach, R Koleňák,Soldering of Copper with High-Temperature Zn-Based Solders, Pro Eng.100 (2015) 1370-1375. https://doi.org/10.1016/j.proeng.2015.01.505

[12] FW Gayle, G Becka, A Syed, J Badgett, G Whitten , TY Pan, A Grusd, B Bauer, R Lathrop, J Slattery, I Anderson, High temperature lead-free solder for microelectronics, Jom. 53 (2001) 17-21. https://doi.org/10.1007/s11837-001-0097-5

[13] JH Bae, K Shin, JH Lee, MY Kim, CW Yang, Development of high-temperature solders: contribution of transmission electron microscopy,App.Mic.45 (2) (2015) 89-94. https://doi.org/10.9729/AM.2015.45.2.89

[14] V Chidambaram, Development of lead-free solders for high-temperature applications, Tec. Uni. Den, (2010).

[15] C Handwerker, U Kattner, KW Moon, Fundamental properties of Pb-free solder alloys,Lea-Fre.Sol. (2007) 21-74. https://doi.org/10.1007/978-0-387-68422-2_2

[16] Miric, Anton-Zoran,New developments in high-temperature, high-performance lead-free solder alloys,Bal.90 (2010) 91-6.

[17] EA Eid, A Fawzy, MM Mansour, G Saad, M Amin,The role of Ni minor additions on the mechanical characteristics of Sn-1.5 Ag-0.5 wt.% Cu (SAC155) Pb-free solder alloy, J. Mat.Sci: Mat.Ele. 35(32) (2024) 2092. https://doi.org/10.1007/s10854-024-13876-8

[18] H Kang, SH Rajendran, JP Jung, Low melting temperature Sn-Bi solder: effect of alloying and nanoparticle addition on the microstructural, thermal, interfacial bonding, and mechanical

Materials Joining and Manufacturing Processes: MJMP 2025   Materials Research Forum LLC
Materials Research Proceedings 55 (2025) 136-140   https://doi.org/10.21741/9781644903612-20

characteristics,Met.11(2) (2021) 364.https://doi.org/10.3390/met11020364
https://doi.org/10.3390/met11020364

[19] Z Moser, J Dutkiewicz, W Gasior, J Salawa, The Sn− Zn (tin-zinc) system,Bul.All.Pha.Dia. 6(4) (1985)330-334. https://doi.org/10.1007/BF02880511

[20] Satyanarayan,KN Prabhu,Reactive wetting of Sn-2.5 Ag-0.5 Cu solder on copper and silver coated copper substrates, J.mat. sci: mat. Ele.24 (5) (2013) 1714-1719.
https://doi.org/10.1007/s10854-012-1002-4

Materials Joining and Manufacturing Processes: MJMP 2025
Materials Research Proceedings 55 (2025) 141-147

Materials Research Forum LLC
https://doi.org/10.21741/9781644903612-21

Thermal spray coatings in industrial boiler environments: A review

Jayson Anil Pinto[1,a*] and K. Narayan Prabhu[1,b]

[1]National Institute of Technology Karnataka, Surathkal, PO Srinivasnagar, Mangalore, Karnataka, India & [1]Mangalore Refinery & Petrochemicals Limited, Mangalore, Karnataka, India

[a]jayson.213ml501@nitk.edu.in, [b]knprabhu@nitk.edu.in

Keywords: Fired Boilers, Corrosion, Thermal Spray Coating

Abstract. Fired Boilers are integral parts of modern Petroleum Refineries and Thermal Power Plants where their reliability directly impacts both the productivity and financial performance. The pressurized components of these boilers are susceptible to various corrosion mechanisms, which can affect the reliability of the boilers. Over the years, a range of corrosion mitigation strategies have been explored and implemented across the industry, yielding varying levels of success. Based on extensive research and field trials, thermal spray coatings have emerged as a promising solution, offering cost-effective corrosion protection. This paper offers a comprehensive overview of thermal spray coatings as an effective strategy for corrosion prevention in industrial boiler environments. Thermal conductivity considerations in the selection of coating materials is highlighted.

Introduction

Steam is one of the most important utility in any petroleum refinery and the main process fluid in a thermal power plant. The entire steam requirement in both cases is derived from Fired Boilers and therefore, their reliability has a direct impact on the productivity and financials of these industrial plants. The pressurized components of Fired Boilers are affected by various damage mechanisms, primarily in the furnace section. Various preventive measures have been adopted to mitigate the damage mechanisms. Provision of thermal spray coatings on pressurized components has been extensively researched and experimented with varying levels of success.

Corrosion & Failure Description

Modern petroleum refineries require steam for their day-to-day operations for motive, heating and process purposes. Steam can constitute upto ~30% of the energy used in the refinery and is produced in fired boilers wherein heat generated by burning fuel oil/gas is used to convert water contained in metallic tubes into steam. Fired boilers are typically made of "water walls" i.e. metallic tubes welded laterally to constitute a metallic tube-wall. The four water walls enclose the furnace wherein heat generated by burning fuel is absorbed by the water flowing through the tubes to generate steam. Although different materials are used in the construction of boiler components, plain Carbon Steel is the most common material of construction.

As in any other industry, corrosion damage and failures are inevitable in fired boilers, with damage mechanisms typically classified based on the initiation surface, as either water-side or fire-side. Some of the common waterside damage mechanisms are Caustic Corrosion, Hydrogen Damage, Overheating (short-term or long-term), Flow Accelerated Corrosion, Stress Corrosion Cracking, Erosion-corrosion etc.

Hydrogen damage is one of the most common failure mechanisms in the water wall section and manifests as a window blow-out or a thick wall rupture. Corrosion reactions at the internal surface result in the production of atomic hydrogen which then diffuses into the steel. The diffused hydrogen either combines at grain boundaries or inclusions to form molecular hydrogen, or reacts with iron carbides to produce methane. Since both molecular hydrogen and methane cannot diffuse

Content from this work may be used under the terms of the Creative Commons Attribution 3.0 license. Any further distribution of this work must maintain attribution to the author(s) and the title of the work, journal citation and DOI. Published under license by Materials Research Forum LLC.

Materials Joining and Manufacturing Processes: MJMP 2025   Materials Research Forum LLC
Materials Research Proceedings 55 (2025) 141-147   https://doi.org/10.21741/9781644903612-21

through the steel, they accumulate in the steel, primarily at the grain boundaries. As gas pressure builds, it causes separation within the metal, leading to the formation of intergranular microcracks. As these microcracks accumulate, the tube's strength decreases to the point until the stresses developed in the tube from boiler pressure surpass the tensile strength of the remaining intact metal, at which stage a thick-walled longitudinal rupture can occur. Depending on the extent of hydrogen damage, large sections of the tube wall may be blown out in the form of a window.

The corrosion reactions that lead to hydrogen generation are either due to low-pH (acid corrosion reactions) or high-pH (caustic corrosion reactions) and usually occur under deposits or beneath corrosion product layers. The tendency to form deposits is influenced by localized heat input, water turbulence, and water composition at or near the tube wall. Deposition tendency is significantly greater at locations where departure from nucleate boiling (DNB) occurs. Nucleate boiling refers to a condition in which discrete bubbles of steam nucleate at points on a metal surface. With increasing heat input, a departure from nucleate boiling (DNB) condition is produced wherein a stable steam layer forms on the metal surface. Unlike nucleate boiling conditions, wherein dissolution of deposits occur, DNB conditions can cause concentration and deposition of even highly water-soluble material.

High heat flux regions favor DNB conditions. In case of compact boilers, heat flux is generally on the higher side. In addition, flame impingement can aggravate the issue. If high heat flux leading to deposition, hydrogen damage and failure is observed in an operating boiler, there is very little that can done to mitigate this issue as the boiler is already designed and constructed. Any alternate solution that can reduce the heat flux will lead to reduced deposition and ultimately untimely failures.

Provision of a coating on the hot/flame exposed side of the water wall tubes can potentially reduce the heat flux at the water-cooled side of the water wall tubes and may therefore lead to reduction in DNB occurrences. In coal fired boilers, weld overlays have been provided on the hot/flame exposed side of the tubes, primarily to combat erosion corrosion. Provision of coatings on the hot/flame exposed side of the tubes has not been explored from the point of view of reducing heat flux.

Impact of Corrosion on Boiler Components
The in-service degradation of boiler steel components may adversely affect their service life, safety and performance while resulting in unplanned breakdowns and maintenance [1]. At thermal power plants in Russia, repair costs account for 12% of the total cost of the electricity generated [2].

High temperature corrosion continues to be one of the main material problems in coal-fired plants and can occur on the steam side (on the internal surface) of the superheater & reheater tubes and on the fireside (on the outer diameter) of the boiler tubes. Corrosion on the fireside depends on the ash, S and Cl present in the coal being used. [3].

Need for Coatings
Though complete elimination of corrosion is impossible, corrosion and its impact can be significantly reduced by deploying effective corrosion control strategies. Historical efforts to tackle corrosion, specifically high temperature corrosion was focused mainly on development of new corrosion resistant alloys which later evolved into standard materials-of-choice for Boiler components.

The growing use of biomass/waste fuels and the push for higher operational efficiency have driven the need for improved corrosion protection in boilers. High operational limits require high performance alloys which tend to be cost prohibitive while having their own unique set of challenges.

Currently, the most effective and cost-efficient corrosion protection method for boiler components is overlay coating, wherein a thin layer of corrosion resistant alloy is provided on a

Materials Joining and Manufacturing Processes: MJMP 2025
Materials Research Proceedings 55 (2025) 141-147

Materials Research Forum LLC
https://doi.org/10.21741/9781644903612-21

substrate of lower corrosion resistant alloy. This approach allows cheaper non-corrosion resistant materials to maintain their mechanical properties while being protected in challenging, corrosive environments and being highly cost-effective at the same time. Several coating processes, such as weld overlay, laser cladding, and thermal spraying, are available, each with their own unique benefits and challenges.

Weld overlays, are corrosion resistant alloys which are directly welded onto the substrate to form a durable, corrosion-resistant metallurgical bond. Weld overlay tend to have issues such as embrittlement, cracking, and dilution of alloying elements which can reduce the effectiveness of the weld overlay. Additionally, the rough surface of the weld overlay can accelerate corrosion, while being susceptible to thermal fatigue cracking of the coated tubes [4].

Laser cladding, which fuses a consumable material onto the substrate using a focused laser beam, produces a dense, crack-free coating. This method minimizes substrate distortion and dilution, offering advantages in precision and speed. However, the benefits are overshadowed by its high cost which limits its adoption in the boiler industry.

In thermal spraying, metallic or non-metallic materials are deposited onto a substrate. The feedstock material, often in powder, wire, or solution form, is heated to a molten or semi-molten state and accelerated in a flame before being sprayed onto the substrate. The molten particles deform upon impact, forming "splats" that bond to the substrate through mechanical interlocking. This process allows for a wide range of materials to be deposited.

To mitigate high temperature degradation of boiler steels, different types of coatings can be successfully employed in different working conditions. However the life and performance of the coating would depend on several factors such as the working environment, the coating process, coating material and the base material. The most commonly used coatings for corrosion and erosion resistance in boiler environments are based on Ni, Cr, Si, Al, WC, Ti and Co. By preventing corrosive species from penetrating the coating, the carbides and oxides of these elements limit the deterioration of the base material.

Thermal Spray Techniques
A brief description of various thermal spray techniques is given below.
Flame Spraying

Flame spraying involves combusting a gasceous fuel such acetylene or propane to generate a flame for melting feedstock materials, which are then sprayed onto a substrate. The process is versatile, portable, and cost-effective but tends to produce coatings with high porosity, limiting its corrosion protection effectiveness.
Detonation Spray

Detonation spray utilises detonation wave to spray coating particles at high temperature of ~4000 °C and velocity of ~1200 m/s. Owing to the high temperature, pressure and velocity, thermal spray coatings deposited by the detonation method have a low porosity and high adhesion to the substrate.
Electric Arc Spray (Wire Arc)

In the electric arc spray method, an electric arc melts consumable electrode wires, and the molten droplets are propelled onto the substrate by compressed air or inert gas. This method offers higher bonding strength, lower porosity, and faster spray rates than flame spraying.

Atmospheric Plasma Spray (APS)

Atmospheric plasma spraying (APS) uses a plasma jet to melt and spray powdered feedstock materials onto a surface. It can spray a wide range of materials including metals and ceramics. The coating is dense and uniform, and is resistant to wear, corrosion, and thermal stress.
High-Velocity Oxygen Fuel (HVOF)

Materials Joining and Manufacturing Processes: MJMP 2025   Materials Research Forum LLC
Materials Research Proceedings 55 (2025) 141-147   https://doi.org/10.21741/9781644903612-21

High-Velocity Oxygen Fuel uses a high-temperature, supersonic gas stream to apply coatings onto surfaces. The resultant coating is dense, has low porosity and high bond strength, providing superior corrosion resistance [5].

High-Velocity Air Fuel (HVAF)

High-Velocity Air Fuel technique is similar to the HVOF process but uses Air instead of Oxygen and lower temperatures than HVOF (just lower than melting point). The use of air, reduces the oxide content of coatings. In addition, the lower flame temperature results in lower thermal degradation of feedstock materials. The combination of lower temperature and lower oxygen content produces coatings with excellent adhesion and low oxide content while preserving protective elements like Cr and Al, necessary for high corrosion resistance.

Cold Spray

In the Cold Spray process, powder particles are accelerated to very high velocities using a supersonic compressed gas jet at temperatures below their melting point. High particle velocities allow intimate conformal contact with the substrate facilitating bonding and enabling the rapid build-up of thick layers of deposited material. Lower process temperatures help to avoid oxidation of deposited material resulting in better coating performance.

Corrosion of Thermal Spray Coatings

Microstructural characteristics and coating chemistry play key roles in corrosion resistance, particularly in high-temperature boiler environments. Corrosion protection of thermal spray coatings depend on the composition, microstructure, and architecture of coatings [4].

Alloying elements primarily influence the corrosion mechanism, while microstructural features affect the corrosion performance. Additionally, coating architecture can improve corrosion protection.

Chromium is a common alloying element in materials for high temperature applications and forms the stable protective oxide Cr_2O_3. The oxide scale is protective in nature as diffusion of corrosive species through it is very slow. The corrosion rate of the substrate follows a parabolic curve, as long as the oxide scale is well attached to the substrate since scale growth is controlled by ion diffusion through the scale [6].

Nickel based alloys are commonly used in high temperature environments. Nickel based alloys perform well in reducing environments. Nickel-based alloys, when alloyed with chromium, exhibit resistance to oxidation, offering a wide range of alloys for optimal corrosion resistance in both oxidizing and reducing conditions.

Aluminum forms a single thermodynamically stable oxide, Al_2O_3. Compared to oxides of other alloys, Al_2O_3 has superior corrosion resistant properties due to its slower growth rate. Although a high Al content in the coating enhances its high-temperature protection, it may degrade the mechanical properties. In an Al_2O_3-forming alloy, the addition of Chromium (Cr) will decrease the amount of aluminum (Al) required to form a protective oxide layer [7].

Si provides corrosion resistance at high temperatures due to the formation of a protective SiO_2 layer. In addition, presence of Si also promotes the formation of other protective oxides such as Cr_2O_3 and Al_2O_3 [8]. Si is typically added in low concentrations as higher concentrations (>3 wt%) have been observed to result in material embrittlement. In addition, under thermal cycling conditions, poor oxide adhesion has been reported in SiO_2-forming alloys [9].

The microstructural features such as pores and splat boundaries in a thermal spray coating affects their corrosion performance. These features can be adjusted through spray process selection, parameters, or post-treatment.

Previous studies on coatings from techniques like APS and HVOF have highlighted issues such as discontinuous oxide scales, pores, and poor splat cohesion, which hinder oxidation resistance. High amounts of interconnected pores and oxide formation at splat boundaries can reduce coating effectiveness. More cohesive, pore-free coatings with less in-situ oxidation could offer better

Materials Joining and Manufacturing Processes: MJMP 2025
Materials Research Proceedings 55 (2025) 141-147

Materials Research Forum LLC
https://doi.org/10.21741/9781644903612-21

performance in high-temperature corrosive environments. Additionally, multi-layered or functionally graded coatings, or composite structures with oxygen-active elements, can improve protection.

To address the typical issues with conventional thermal spray coating processes, new processes have been developed such as the nano-structured coating process and the cold spray coating process. The nano-structured coating process offers coatings with high-quality surface finish and lower porosity while the cold spray coating technique mitigates the oxidation issues associated with high temperature coating processes.

Additionally, to significantly improve the quality, performance and life of thermal spray coatings, post coating treatments have been developed such as:

Laser re-melting of the thermal spray coating leading to decreased porosity, uniform microstructure and increased corrosion resistance [10];

Heat treatment to compact microstructure, increase splat bonding, corrosion and wear resistance [11]

Thermal Spray Coatings for Industrial Fired Boilers

Several thermal spray techniques have been used to evaluate coating systems for industrial boiler applications such as Plasma Spray, Detonation Spray, Electric Arc Spray, High Velocity Oxygen Spray and Cold Spray [1].

Several Fe based alloys are used in industrial boilers components such as Plain Carbon Steels, Low Alloy Steels (1.25Cr-0.5Mo, 2.25Cr-1Mo, 9Cr-1Mo, 9Cr-1Mo-V) and Stainless Steels (304, 316, 310, 321, 347 & 410). Selection of steel is dependent on the operating temperature and internal pressure of the component. Plain carbon steels are used at low temperatures such as for the boiler feed water piping, the water wall tubes and the economizer piping. The alloy steels are used in steam service from moderate to high temperature and the stainless steels are used for high temperature applications with high radiant heat flux.

Literature review indicates extensive research has been carried out on boiler steels coated with several coating systems. A summary of the current literature on the subject is provided below:

9Cr-1Mo steels coated with Ni-Cr, Cr_3C_2-NiCr, NiCrAl and Cr- & Ni- based Fe coatings (deposited by the HVOF and Plasma spray processes) were evaluated for high temperature corrosion and found suitable [1].

Significant research effort has been undertaken on investigating the performance of thermal spray coatings on 2.25Cr-1Mo steels. Studies have explored a range of coating systems applied to these alloys, evaluating their effectiveness in various conditions. Cr_3C_2-NiCr coatings were evaluated for high temperature oxidation and high temperature erosion corrosion and found to be suitable. Performance of Cr_2O_3 and Cr_2O_3-Al_2O_3 coatings were found to be suitable. The WC-12Co coating was found to be non-effective for high temperature oxidation while the Ni-20Cr coating was found to be 100% more corrosion resistant than the uncoated sample. 80Ni-20Cr and 50Cr-50Ni were evaluated for high temperature oxidation and the 50Cr-50Ni was found to be more effective at resisting oxidation [1].

SS304 samples coated with NiCrSiB-Al_2O_3 and NiCrSiB-75Cr_3C_2-25NiCr systems were evaluated for erosion resistance and were found to provide significant resistance to erosion corrosion. Ni-Al coating on SS304 were not found beneficial in providing high temperature corrosion resistance. SS316 provided with WC-12Co and WC-10Co-4Cr coatings performed well in erosion and abrasion wear tests than Cr_3C_2-25NiCr coatings. Ni_3Al coatings on SS321 were not found effective due to internal oxidation of the coating itself. Ni-20Cr coating was found effective at resisting high temperature corrosion as were Cr_3C_2-NiCr coatings on SS347H [1].

Thermal Conductivity Considerations in Thermal Spray Coating Systems

Physical properties of thermal spray coatings depend on several factors such as coating microstructure, feedstock material, oxide, impurities, and contaminations [12]. In the thermal spray process, feedstock materials are partially or completely melted, and the particles are rapidly transported to the substrate, where they quickly solidify upon impact to form "splats". When these splats are built up one after the other, the coating becomes layered, resembling a brick wall with intricate arrays of interwoven splats. As a result of this splat-based multilayer microstructure, the coating exhibits inherent anisotropy in the direction perpendicular to the spray direction. In other words, coating property anisotropy is a result of microstructural anisotropy [13].

The thermal conductivity of a thermal spray coating is highly influenced by both the feedstock material and the process parameters. In many instances, the thermal properties of the coating can differ significantly from those of the corresponding bulk material [14], particularly in the case of metal coatings [15]. Thermal diffusivity of thermal spray coatings obtained by the laser flash technique were observed to be much lower than that of the bulk material [16]. The lower thermal conductivity of thermal sprayed coating than the bulk material with the same composition may be due to the contribution of microstructural features (pores, cracks, etc.).

Thermal conductivities of thermal spray coating materials are dependent on the temperature and chemical composition. The inter-lamellar porosity plays an important role in both room-temperature and temperature-dependent thermal conductivity [17, 18]. Experimental results indicate an inverse relationship between thermal conductivity and total porosity.

The coating process parameters have a significant impact on the resultant coating properties. For example, changes in temperature and velocity settings during coating can cause a substantial variation in total porosity (upto 10%), due to differences in the melting and kinetic states. Higher temperatures and velocities lead to a higher flattening ratio of splats, higher interfaces per unit thickness, enhanced splat contact and decreased inter-lamellar porosity. As a result, the variation in total porosity caused by different particle states has a significant impact on variation in thermal conductivity.

Summary

Corrosion of pressurized boiler steel components is an existential threat significantly impacting dependent industrial units. Extensive research has been conducted to mitigate this issue through thermal spray coatings. Various coating systems have been developed and evaluated, demonstrating strong potential for industrial application. Research has been primarily focused on increasing the life of boiler tubes using thermal spray coatings. The impact of thermal conductivity on heat transfer and consequential influence on corrosion has not been studied till date. Thermal spray coatings on the outer surface of the boiler tubes can potentially reduce the heat flux at the inner, water-cooled side thereby decreasing the likelihood of DNB occurrences and resultant tube failures. The selection of spray coatings for preventing corrosion should be based on assessment of heat flux at the coating/ambient interface.

References

[1] D. Dhand, P. Kumar, and J. S. Grewal, A review of thermal spray coatings for protection of steels from degradation in coal fired power plants, Cor. Rev. 39 (2021) 243-268. https://doi.org/10.1515/corrrev-2020-0043

[2] E. Kakaras, Current situation of coal fired power plants in Russian Federation and the implementation options of clean coal technologies, Proceedings of the 5th European Conference on Coal Research & Its Applications, Edinburgh, UK, (2004).

[3] B. A. Pint, High-temperature corrosion in fossil fuel power generation: present and future, JOM 65 (2013) 1024-1032. https://doi.org/10.1007/s11837-013-0642-z

[4] E. Sadeghi, N. Markocsan, and S. Joshi, Advances in corrosion-resistant thermal spray coatings for renewable energy power plants. Part I: Effect of composition and microstructure, J. Ther. Spr. Tech. 28 (2019) 1749-1788. https://doi.org/10.1007/s11666-019-00938-1

[5] S. Kuroda, J. Kawakita, M. Watanabe, and H. Katanoda, Warm spraying - A novel coating process based on high-velocity impact of solid particles, Sci. Tech. Adv. Mat. 9(3) (2008). https://doi.org/10.1088/1468-6996/9/3/033002

[6] A. Zahs, M. Spiegel, and H. J. Grabke, The influence of alloying elements on the chlorine-induced high temperature corrosion of Fe-Cr alloys in oxidizing atmospheres, Mat. & Corr. 50 (1999) 561-578. https://doi.org/10.1002/(SICI)1521-4176(199910)50:10<561::AID-MACO561>3.0.CO;2-L

[7] N. Israelsson, High Temperature Oxidation and Chlorination of FeCrAl Alloys, Doctoral Thesis, Chalmers University of Technology, (2014).

[8] H. J. Grabke, M. Spiegel, and A. Zahs, Role of alloying elements and carbides in the chlorine-induced corrosion of steels and alloys, Mat. Res. 7(1) (2004) 89-95. https://doi.org/10.1590/S1516-14392004000100013

[9] P. Viklund, High Temperature Corrosion during Waste Incineration Characterization, Causes and Prevention of Chlorine-Induced Corrosion, Licentiate Thesis, KTH Royal Institute of Technology, (2011).

[10] M. Buchtík et al., Influence of laser remelting on the microstructure and corrosion behavior of HVOF-sprayed Fe-based coatings on magnesium alloy, Mat. Char. 194 (2022) 112343. https://doi.org/10.1016/j.matchar.2022.112343

[11] J. A. Morales, O. Piamba, J. Olaya, and F. Vallejo, Effect of heat treatment on the electrochemical and tribological properties of aluminum-bronze coatings deposited using the thermal spraying process, Coat. 14 (4) (2024). https://doi.org/10.3390/coatings14040423

[12] Y. Tan, J. P. Longtin, and S. Sampath, Modeling thermal conductivity of thermal spray coatings: comparing predictions to experiments, J. Ther. Spr. Tech. 15 (2006) 545-552. https://doi.org/10.1361/105996306X147216

[13] I. Sevostianov and M. Kachanov, Anisotropic thermal conductivities of plasma-sprayed thermal barrier coatings in relation to the microstructure, J. Ther. Spr. Tech. 9 (2000) 478-482. https://doi.org/10.1007/BF02608549

[14] R. McPherson, A model for the thermal-conductivity of plasma-sprayed ceramic coatings, Thin Sol. Fil. (1984) 89-95. https://doi.org/10.1016/0040-6090(84)90506-6

[15] S. Sampath, X. Y. Jiang, J. Matejicek, A. C. Leger, and A. Vardelle, Substrate temperature effects on splat formation, microstructure development and properties of plasma sprayed coatings: Part I. Case study for Zirconia, Mat. Sci. Eng. A (1999) 181-188. https://doi.org/10.1016/S0921-5093(99)00459-1

[16] J. F. Lagerdrost and K. E. Wilkes, Thermophysical properties of plasma sprayed coatings CR-121144, Nat. Aer. & Spa. Adm. (NASA) (1973).

[17] L. Pawlowski and P. Fauchais, Thermal transport properties of thermally sprayed coatings, Int. Mat. Rev. 37 (1) (1992) 271-290. https://doi.org/10.1179/imr.1992.37.1.271

[18] K. W. Schlichting, N. P. Padture, and P. G. Klemens, Thermal conductivity of dense and porous Yttria stabilized Zirconia, J. Mat. Sci. 36 (2001) 3003-3010. https://doi.org/10.1023/A:1017970924312

Materials Joining and Manufacturing Processes: MJMP 2025
Materials Research Proceedings 55 (2025) 148-155

Materials Research Forum LLC
https://doi.org/10.21741/9781644903612-22

Characterization and thermal performance of Sn-Bi alloy used as a thermal interface material

Kumar Swamy M.C.[1,a*], Satyanarayan[1,b]

[1]Department of Mechanical Engineering, Alva's Institute of Engineering and Technology Karnataka, Moodbidri, Mangalore 574225 and affiliated to Visvesvaraya Technological University, Belagavi, Karnataka, India

[a]mckswamy@aiet.org.in, [b]satyan.nitk@gmail.com

Keywords: Thermal Interface Material (TIM), Thermal Contact Resistance (TCR), Micro Hardness, Micro Structure

Abstract. The growing demand for efficient heat dissipation in electronic devices necessitates the development of high-performance TIMs with superior thermal conductivity and mechanical stability. Sn-Bi alloys, known for their low melting points and good wettability, offer alternatives to conventional TIMs. The present study focuses on the microstructural studies, thermal conductivity, hardness and interfacial resistance of Sn-40Bi (alloy 1), Bi-42Sn (alloy 2) and Bi-30.8Sn-29.8Pb (alloy 3) for thermal management applications. Different thicknesses of Sn-Bi and Bi-Sn-Pb alloys were preferred and further, the results are compared. Differential scanning calorimetry (DSC), and scanning electron microscopy (SEM), were used to characterise the alloys. The findings provide insights into the feasibility of Sn-Bi alloys as efficient TIMs. TIM specimen of thickness 0.5 mm exhibited lowest value of TCR for all the alloys. Sn–40Bi thermal interface alloy exhibited better thermal performance compared to Bi–42Sn and Bi–30.8Sn–29.8Pb.

Introduction

Thermal management plays a very important role in the electronics component cooling system because conventional electronic circuits built by meagre devices such as resistors, transistors, inductors and capacitor exhibit limited functionality. Heat management is a crucial factor for optimal performance, enhanced stability, and extended service life of electronic devices [1–2]. If the system permits, generated heat can be dissipated outside through an improved heat sink, increased air velocities, and liquid cooling. [3] A substantial heat sink employing an advanced cooling method is preferable as interface material. Diverse types of interfaces exist between heat-generating components and heat sinks. In certain instances, a microscale rough surface may result in a minimal contact area between the two surfaces. Therefore, thermal interface materials (TIM) are necessary to improve surface contact and diminish thermal interfacial resistance. The purpose of TIMs is to occupy the microscale voids between two contacting materials to improve thermal conduction at the interfaces. It is evident that two solid materials rarely make direct contact when brought together; significant thermal contact resistance (TCR) arises from the minute roughness of heat sinks and integrated circuits. [4] Over the decades, electronic cooling package design experts have utilized many categories of TIMs. They comprise grease, gel, Phase Change Material (PCM), thermal pads, adhesives, solders, and carbon-based TIMs, among others. [5-7] Thermal grease, gel, and phase change materials shown commendable performance qualities during operation; however, a significant downside is pump-out, which adversely impacts the thermal resistance between the materials. According to the survey, optimal TIMs are characterized by low resistance, high thermal conductivity, and sufficient surface contact. To enhance thermal performance, TIMs such as low-melting-point alloys (LMAs) are utilized in electronic cooling package systems. [8-9] LMAs possess superior thermal conductivity relative to conventional TIMs and exhibit exceptionally low thermal resistance. [10]

Content from this work may be used under the terms of the Creative Commons Attribution 3.0 license. Any further distribution of this work must maintain attribution to the author(s) and the title of the work, journal citation and DOI. Published under license by Materials Research Forum LLC.

Materials Joining and Manufacturing Processes: MJMP 2025   Materials Research Forum LLC
Materials Research Proceedings 55 (2025) 148-155   https://doi.org/10.21741/9781644903612-22

This study examines the microstructural characteristics, thermal conductivity, hardness, and interfacial resistance of Sn-40Bi, Bi-42Sn, and Bi-30.8Sn-29.8Pb for thermal management applications. Various thicknesses of Sn-Bi alloys were selected, and the results were subsequently compared. Differential scanning calorimetry (DSC) and scanning electron microscopy (SEM) were employed to characterize the alloys.

Material And Properties

LMAs were obtained in the form of ingots from Roto Metals Inc. (a US-based company that specializes in a wide range of non-ferrous metals and alloys). The properties of these alloys are listed in Table.1.

Table .1 *Characteristics of the LMAs*

Sl. No	Alloy Name	Composition (% by mass)	Melting Point (°C)	Density $\rho(Kg/m^3)$	Specific Heat $C_p(J/Kg\ K)$
1	Roto281-338F	Sn–40Bi	133° C to 170° C	7350	184
2	Lead–free, Fishing Tackle Weight Bi-Tin Alloy - 281	Bi–42Sn	133° C	7979	167
3	Roto212F Bismuth Based Alloy	Bi–30.8Sn–29.8Pb	100°C	8704	157.04

Thermal conductivity is a crucial property of TIMs that has a direct result on the total interfacial resistance. Measurement of thermal diffusivity and thermal conductivity were conducted at Department of Mechanical Engineering, IIT Kanpur. Table 2 provides the thermal conductivity values of TIMs used in the study.

Table.2 *Thermal diffusivity and conductivity of purchased alloys*

Sl. No	Alloy	Thermal Diffusivity(mm^2/s)	Thermal conductivity (W/mK)
1	Sn-40Bi	19.8	26.6
2	Bi-42Sn	14.9	19.8
3	Bi-30.8Sn-29.8Pb	12.5	17.1

Fig.1 illustrates the results derived from the micro hardness tests conducted on the TIM alloy. Sn-40Bi exhibited lowest hardness whereas the Bi-Sn–Pb alloy the highest. An increase in the hardness was attributed to the existence of Pb. Lead is known to increase the hardness of tin-based alloys, such as pewter alloy, by acting as a solid solution strengthening element. When Pb atoms are introduced into the crystal lattice of the alloy, they hinder the movement of displacements, which are line imperfections in the crystal structure responsible for plastic deformation. This hindrance strengthens the material and increases its hardness [11].

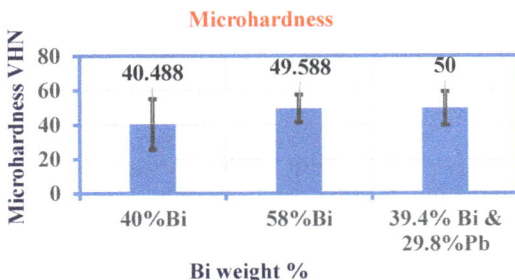

Fig. 1 *Micro hardness of all the three alloys*

Experimental Procedure:

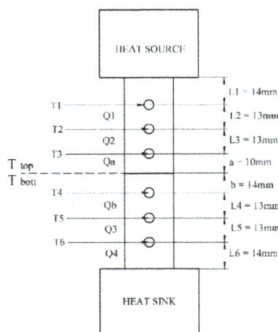

Fig.2 *Schematic of Test Rig and thermocouple arrangement*

To study the thermal performance of alloys, widely accepted the ASTM D-5470 standardized experimental setup method was adopted. A schematic representation of the actual test configuration is illustrated in Fig. 2. The setup comprises two solid bars constructed from Oxygen-Free High Conductivity (OFHC) copper having thermal conductivity 400 W/m·K. The copper blocks measure 50 mm in length and include a round cross-section with a diameter of 25 mm. The opposing surfaces were meticulously honed to a level finish. The upper bar is heated electrically by embedded resistance heaters. The temperature distribution in each rod was measured using three evenly spaced K-type thermocouples, which were inserted in equally spaced drills (∅ 0.5 mm diameter) 13 mm apart into both Cu rods. The current investigation employed the steady state approach owing to a reduced percentage of experimental uncertainty.

Result and Discussion

Fig. 3 shows the DSC curves obtained by adopting heating rate of 10°C per minute from the temperatures of 20°C to 400°C. Compared with Sn-40Bi alloy, and Bi-42Sn, Bi–30.8Sn-29.8Pb alloy has involved in lower heat absorption which undergoes phase transformation relatively at lower temperature than Sn-40Bi alloy. The reason is that Sn-40Bi possess higher heat capacity and thermal conductivity compared to other two alloys. The solidus temperatures, liquidus temperatures and peak temperatures of three alloys are tabulated in Table. 3.

Materials Joining and Manufacturing Processes: MJMP 2025   Materials Research Forum LLC
Materials Research Proceedings 55 (2025) 148-155   https://doi.org/10.21741/9781644903612-22

***Table.* 3** *DSC analysis report*

Sl.NO	Composition	Solidus Temperatures	Liquidus Temperatures	Peak Temperature
1	Sn-40Bi	143.5°C	170°C	156.4°C
2	Bi-42Sn	142.8° C	168°C	154.7°C
3	Bi-30.8Sn-29.8Pb	106.7° C	130° C	112.1° C

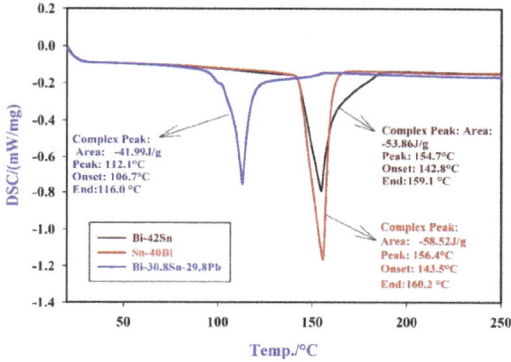

***Fig.* 3** *DSC curve of LMAs*

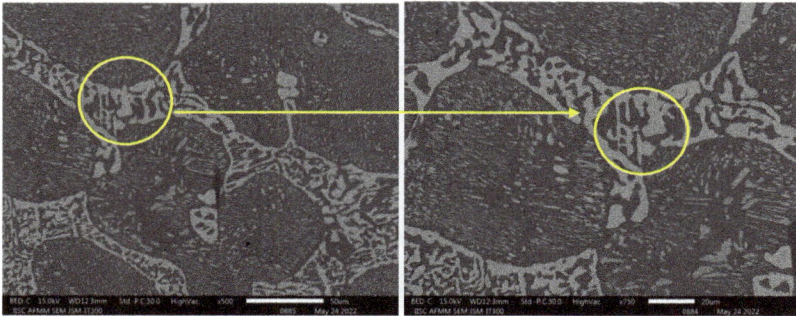

Fig.4 *SEM images of Sn-40Bi alloy*

Fig. 4 shows back scattered SEM (BSE) images of Sn-40Bi alloy. The images revealed a microstructure consisting of Bi-rich and Sn-rich phases, represented by the bright and dark grey regions, respectively. The alloy was clearly identified as having a laminar Sn-Bi eutectic structure and Sn dendrites with Bi particles/strips. The fine-grained microstructure of Sn-40Bi alloy is particularly important for its thermal properties. The near eutectic composition of the alloy results in a microstructure with alternating bands of Sn and Bi phases, which form a network of interconnected channels. These channels provide a pathway for heat transfer through the material, and the fine scale of the microstructure allows for efficient thermal transport.

151

Materials Joining and Manufacturing Processes: MJMP 2025   Materials Research Forum LLC
Materials Research Proceedings 55 (2025) 148-155   https://doi.org/10.21741/9781644903612-22

SEM images of Bi-42Sn alloy are shown in Fig. 5. The Image revealed that, microstructure was a usually lamellar eutectic structure with alternate layers of white Bi phase and grey Sn phase. Due to the immediate phase separation during the solidification process, its structure reflected a typical striped eutectic one with alternated β-Sn phases and Bi phases. The finer eutectic can be seen as individual "islands" located between the eutectic mixture and the Sn-rich dendrites. This microstructural configuration belongs to the class of discontinuous eutectic structures known as "Chinese script" structures. In addition, the "fishbone" eutectic structure was observed in SEM image.

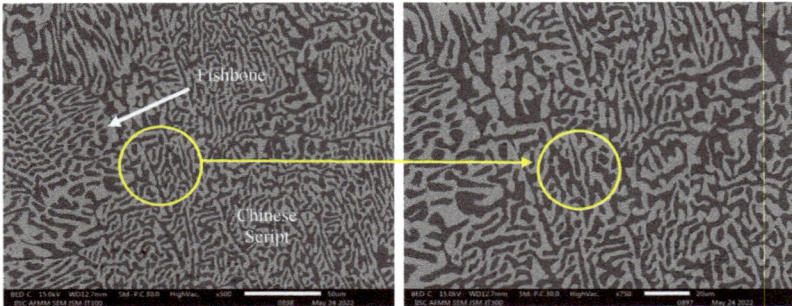

Fig. 5 SEM images of Bi-42Sn alloy

Typically, this structure prefers to appear adjacent to the lamellar structure of Bi-rich phase. The fishbone-like structure in Sn-58Bi specimens appears to be a region distinct from the two distinct scales of lamellar eutectic phases. Compared to Sn-40Bi alloy, Bi-42Sn exhibited coarser and finer microstructure this could be attributed to thermal instability caused by local heat flow. SEM images of Bi-30.8Sn-29.8Pb alloy are shown in Fig. 6. The alloy exhibited a quasi-regular structure. The BSE image revealed two regions: a white region and a black island-shaped region.

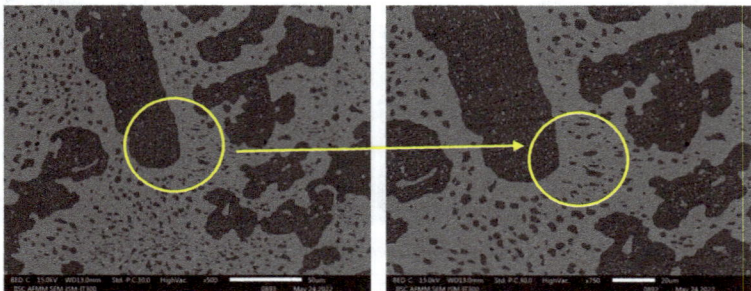

Fig. 6 SEM images of Bi-30.8Sn-29.8Pb alloy

The black island-shaped section was unevenly distributed throughout the white region. The majority of the Pb and Bi components were dispersed in the white region, whereas the majority of the Sn elements were distributed in the black region. Image revealed that, Pb element was not spread uniformly in the Bi-rich region.

Materials Joining and Manufacturing Processes: MJMP 2025   Materials Research Forum LLC
Materials Research Proceedings 55 (2025) 148-155   https://doi.org/10.21741/9781644903612-22

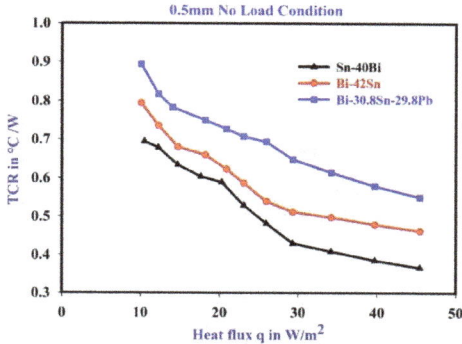

Fig. 7 *Thermal contact resistance v/s heat flux for Sn-40Bi, Bi-42Sn, and Bi-30.8Sn-29.8Pb alloys*

Fig. 7 shows the nature of variation of TCR versus heat flux for the alloys Sn-40Bi, Bi-42Sn, and Bi-30.8Sn-29.8Pb. It provides insight on the performance of alloys in terms of heat flow efficiency and TCR under no load condition. The thermal conductivities of the alloys play a crucial role in determining their heat transfer capabilities. In the study, Sn-40Bi has exhibited the highest thermal conductivity at 26.6 W/mK, whereas Bi-42Sn showed 19.8 W/mK, and Bi-30.8Sn-29.8Pb with a thermal conductivity of 17.1 W/mK. Greater thermal conductivities typically indicate greater heat transfer capabilities, as materials with greater thermal conductivities can conduct and dissipate heat more efficiently. Consequently, alloys with greater thermal conductivities have reduced TCR. From the graph it was noticed that, Sn-40Bi exhibited the lowest TCR on the graph, followed by Bi-42Sn and Bi-30.8Sn-29.8Pb. Since the development of microstructure is dependent on the element utilised in the alloy, the addition of an element will directly affect the aspects of hardness and thermal conductivity[12], [13]. As a result, TIM (Sn-40Bi), with higher thermal conductivity and lower hardness, performed well in terms of heat flux between interfaces.

Fig. 8 shows the nature of the effect of TCR versus thickness for Sn-40Bi, Bi-42Sn, and Bi-30.8Sn-29.8Pb alloys at different thermal conductivities. The TCR was increased as the thickness of the material layer between the surfaces was increased. This is because a thick layer creates a longer path barrier for heat flux, resulting in greater heat transfer resistance. Thus, the thermal contact resistance plot for all the alloys was increased as the thickness was increased. Alloys showed that as the thickness of the material was increased, the thermal contact resistance increased for all three alloys. However, the rate of increase was influenced by the thermal conductivity of the alloy. Alloys with higher thermal conductivity (Sn-40Bi) exhibited a slower increase in thermal contact resistance with thickness, while alloys with lower thermal conductivity (Bi-30.8Sn-29.8Pb) indicated a more significant increase in thermal contact resistance. This relationship emphasises the significance of taking material thickness and thermal conductivity into account when evaluating thermal contact resistance at surfaces.

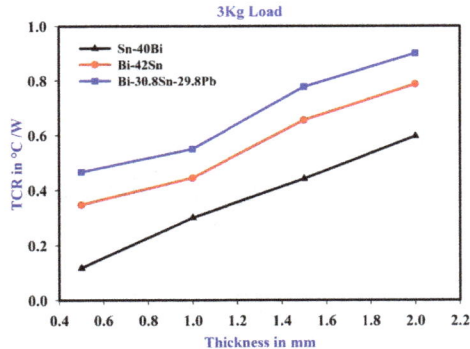

Fig. 8 *Thermal contact resistance v/s thickness for Sn-40Bi, Bi-42Sn, and Bi-30.8Sn-29.8Pb alloys loaded at 3Kg.*

Conclusion

Based on the results and conclusions, the following conclusions are drawn

➢ Microstructures of Sn–40 Bi TIM alloy exhibited a eutectic morphology in which Bi phase particles were uniformly dispersed in the matrix. The white region of Bi phase appeared like an island shape whereas Bi–42Sn alloy appeared as a class of discontinuous eutectic structures called as 'Chinese script' or 'fishbone'. The island shape of Bi appeared unevenly in Bi-30.8Sn-29.8Pb alloy.

➢ Sn–40 Bi exhibited higher thermal conductivity (26.6 W/mK) and thermal diffusivity (19.8 mm/s) in compared to Bi-42Sn (19.8 W/mK, 14.9 mm/s) and Bi-30.8Sn-29.8Pb (17.1 W/mK, 12.5 mm/s)

➢ The micro–hardness of Sn–40Bi alloy was found to be 40.488 HV, whereas for Bi–42Sn and Bi-30.8Sn-29.8Pb alloys the values were observed exhibited to be 49.588 HV and 50 HV.

➢ The thermal contact resistance (TCR) for 0.5 mm thickness of Sn–40Bi TIM alloy under no load condition (0kg) was found to be 0.3660 °C /W, whereas TCR values under the same condition for Sn–40 Bi and Bi–30.8Sn-29.8Pb TIM alloys were found to be 0.462 °C /W and 0.550 °C /W.

➢ Sn–40Bi TIM alloy exhibited TCR of 0.1180 °C /W under full load condition (3kg) whereas for Bi-42Sn and Bi-30.8Sn-29.8Pb TIM alloys the TCR was found to be 0.346 °C /W and 0.465 °C /W

➢ Sn–40Bi TIM alloy exhibited better thermal interface performance than the Bi–42Sn and Bi–30.8Sn–29.8Pb TIM alloys.

References

[1] M.C. Kumar Swamy and Satyanarayan, "A Review of the Performance and Characterization of Conventional and Promising Thermal Interface Materials for Electronic Package Applications", J. Electron. Mater, 48 (12) (2019) 2019. https://doi.org/10.1007/s11664-019-07623-7

[2] M.C. Kumar Swamy and Satyanarayan, "Study on thermal resistance of brass with and without coating of metallic surface", Mater. Today Proc., 35 (2021) 335–339. https://doi.org/10.1007/s11664-019-07623-7.

Materials Joining and Manufacturing Processes: MJMP 2025   Materials Research Forum LLC
Materials Research Proceedings 55 (2025) 148-155   https://doi.org/10.21741/9781644903612-22

[3] V.V. Andra´s, Z. Sa´rka´ny, and M. Rencz, Characterization method for thermal interface materials imitating an in-situ environment, Microelectron. J.43, 661 (2012). https://doi.org/10.1016/j.mejo.2011.06.013

[4] M. Grujicic, C.L. Zhao, E.C. Dusel, The effect of thermal contact resistance on heat management in the electronic packaging, Appl. Surf. Sci. 246 (2005) 290–302.https://doi.org/10.1016/j.apsusc.2004.11.030

[5] J. Due, A. J. Robinson, Reliability of thermal interface materials: A review, Appl. Therm. Eng., 50(1) (2013): 455-463.https://doi.org/10.1016/j.applthermaleng.2012.06.013

[6] W. X. Chu, M. Khatiwada, C. C. Wang, Investigations regarding the influence of soft metal and low melting temperature alloy on thermal contact resistance, Int. Commun. Heat Mass Transfer, 116, p.104626.https://doi.org/10.1016/j.icheatmasstransfer.2020.104626

[7] C.K. Roy, S. Bhavnani, M.C. Hamilton, R.W. Johnson, R.W. Knight, D.K. Harris, Thermal performance of low melting temperature alloys at the interface between dissimilar materials, Appl. Therm. Eng. 99, 72–79 (2016) https://doi.org/10.1016/j.applthermaleng.2016.01.036

[8] C.G. Macris, T.R. Sanderson, R.G. Ebel, C.B. Leyerle, Performance, Reliability, and Approaches Using a Low Melt Alloy as a Thermal Interface Material, Proceedings IMAPS. 2004. Available from: https://enerdynesolutions.com/downloads/imaps_2004_man.pdf

[9] E. Yang, H. Guo, J. Guo, J. Shang, M. Wang, Thermal performance of low-melting-temperature alloy thermal interface materials, Acta Metall. Sin. Engl. Lett. 27 (2) (2014) 290–294https://doi.org/10.1007/s40195-014-0042-6

[10]R. D. Pathumudy, K. N. Prabhu, Thermal interface materials for cooling microelectronic systems: present status and future challenges, J. Mater. Sci. - Mater. Electron. 32 (2021): 11339-11366. https://doi.org/10.1007/s10854-021-05635-w

[11]S. C. Britton, 'Tin and Tin Alloys', Corros. Third Ed., vol. 1, pp. 4:157-4:167, 2013, https://doi.org/10.1016/B978-0-08-052351-4.50048-0

[12]H. R. Kotadia, P. D. Howes, and S. H. Mannan, 'A review: On the development of low melting temperature Pb-free solders', Microelectron. Reliab., vol. 54, no. 6–7, pp. 1253–1273, 2014, https://doi.org/10.1016/j.microrel.2014.02.025

[13]P. Das, S. Bathula, and S. Gollapudi, 'Evaluating the effect of grain size distribution on thermal conductivity of thermoelectric materials', Nano Express, vol. 1, no. 2, 2020, https://doi.org/10.1088/2632-959X/abb43f

Materials Joining and Manufacturing Processes: MJMP 2025   Materials Research Forum LLC
Materials Research Proceedings 55 (2025) 156-161   https://doi.org/10.21741/9781644903612-23

Mechanical performance of 3D-printed PBAT composites reinforced with sawdust

Maruthi Prashanth B.H.[1], P.S. Shivakumar Gouda[2], Sandeepkumar Gowda[1,a], Asif Iqbal Mulla[3], Girish Ariga[4], Suresh P.S.[5], Sandeep Kyatanavar[3] and Srinivasa C.S.[5]

[1]Department of Mechanical Engineering, AGMR College of Engineering and Technology, Hubli, Karnataka India-581207 and Affiliated to Visvesvaraya Technological University

[2]Department of Robotics & AI, MITE, Mangalore -575001

[3]Department of Electronics and Communication, AGMR College of Engineering and Technology, Hubli, Karnataka India-581207 and Affiliated to Visvesvaraya Technological University

[4]Department of Basic Science & Humanities, AGMR College of Engineering and Technology, Hubli, Karnataka India-581207 and Affiliated to Visvesvaraya Technological University

[5]Department of Mechanical Engineering, Alva's Institute of Engineering and Technology Moodbidri, Dakshina Kannada-574225 and Affiliated to Visvesvaraya Technological University

[a]sandeepg0404@gmail.com

Keywords: Polybutylene Adipate Terephthalate, 3D Printing, Sawdust, Mechanical Properties

Abstract: This study examines the mechanical properties of polybutylene adipate terephthalate (PBAT) composites reinforced with sawdust (SD) at varying concentrations (2%, 4%, and 6%), fabricated using 3D printing. The influence of SD content on tensile, flexural, and impact properties was analyzed. The results indicate that incorporating 4% SD yields the best mechanical performance, with a tensile strength of 23 MPa, tensile modulus of 512 MPa, flexural strength of 32 MPa, flexural modulus of 1672 MPa, and impact strength of 98 J/m. However, increasing SD to 6% leads to increased brittleness, reducing tensile strength (21 MPa), tensile modulus (494 MPa), flexural strength (30 MPa), flexural modulus (1459 MPa), and impact strength (89 J/m). These findings demonstrate that SD reinforcement enhances PBAT composites, providing a balance of strength and flexibility, making them suitable for lightweight applications.

Introduction

Traditional plastics like polyethylene, polypropylene, and polystyrene contribute to severe white pollution, driving research on biodegradable polymers to minimize environmental impact [1]. PBAT is a top biodegradable polymer for agriculture, packaging films, and medical devices. PBAT, a flexible and ductile aliphatic-aromatic copolyester from fossil resources, biodegrades completely within weeks through enzymatic action [2]. PBAT's high cost and low stiffness limit its use, but blending it with natural fibers preserves biodegradability and impact resistance, making biocomposites ideal for single-use items like fast-food utensils and food containers [3]. Biodegradable composite fillers reduce costs, enhance mechanics, and add functionality. Biochar in PBAT/PLA blends altered surface resistivity, expanding application potential [4].

Yang et al. [5] enhanced lignin sulfonate nanoparticles with maleic anhydride (MLS) and blended them with PBAT, achieving improved tensile strength, elongation, and modulus at 5% MLS, while reducing viscosity and enhancing processability. Jessica et al. [2] reinforced PBAT with untreated and GPTMS-treated peach palm fibers ("pupunha"), finding reduced tensile

Content from this work may be used under the terms of the Creative Commons Attribution 3.0 license. Any further distribution of this work must maintain attribution to the author(s) and the title of the work, journal citation and DOI. Published under license by Materials Research Forum LLC.

Materials Joining and Manufacturing Processes: MJMP 2025   Materials Research Forum LLC
Materials Research Proceedings 55 (2025) 156-161   https://doi.org/10.21741/9781644903612-23

strength and elongation but significantly improved tensile modulus, especially with treated fibers. Increased fiber content enhanced fiber-matrix interaction, raised glass transition temperature, and lowered tan delta peak height. Arvind et al. [3] used maleic anhydride-grafted PBAT (mPBAT) to enhance the interfacial compatibility of hemp powder (HP) in HP/PBAT biocomposites, achieving a 209% increase in tensile strength, 300% in toughness, 90% in impact resistance, and a 60°C higher heat deflection temperature at 40% HP content. Vinicius et al. [6] produced PBAT/PHB/Babassu composites via melt extrusion and analyzed their thermal, mechanical, and morphological properties. PBAT inhibited PHB crystallization and acted as a plasticizer, while Babassu reduced tensile strength and elongation. Higher PBAT content improved elongation and fracture refinement. Wataya et al. [7] studied blending of PVC with PLA, PBAT, and WF (wood flour) which enhanced its mechanical, thermal, and biodegradable properties. PVC/PBAT had the best impact strength, PVC/PLA excelled in flexural and thermal properties, and WF improved all blends.

The literature review shows that PBAT is commonly used to enhance PLA's mechanical properties, but limited research exists on pure PBAT reinforced with sawdust. In this study, PBAT is coated with 2, 4, and 6 wt.% sawdust using a mixer, and composites are manufactured via 3D printing following ASTM standards. The specimens are tested for tensile, flexural, and impact strength.

Materials & Method

PBAT Ecoflex grade 1200, with a melt flow index of 2.7–4.5 g/10 min (170°C, 2.1 kg), a density of 1.23–1.26 g/cm³, and a melt temperature range of 100–108°C, was sourced from Polynomous Industries Private Limited, Gautam Budh Nagar, Uttar Pradesh, India. Sawdust was purchased from a local sawmill and it has density of 0.65 g/cm³. Procured saw dust (SD) was first washed in water for 5 minutes to remove dirt, then oven-dried at 40°C for an hour and sifted to under 300μm. It was then treated in 5 wt.% NaOH solution for an hour, neutralized with 5% HCl, rinsed with water, and oven-dried at 40°C for another hour. PBAT pallets are then coated with saw dust with different saw dust wt.% (2, 4, 6) in a multi roll coating machine. Ethanol is utilized as a solvent during the coating process, which aids in sawdust particles adhering to the PBAT Pallets. Melting sawdust-coated PBAT pallets is the first step in creating the 3D-printed SD-PBAT composite. Composite composition for different samples is shown in table 1. Tensile testing evaluates material strength and deformation per ASTM D3039 [8] standards (250mm × 25mm × 4mm). Using an Instron 1195 (10 KN), samples (12.7mm × 125mm × 4mm) with a 100mm support span are tested for maximum stress before failure as per ASTM D790 [9]. As per ASTM D256 [10] samples (64mm × 12.7mm × 4mm) were vertically mounted and struck by the impact tester in Zwick-Roell Izod test equipment.

Table 1: Composite Composition

Sl. No.	Saw dust		Epoxy Resin and hardener		Sample Name
	% by Weight	Weight in grams	% by Weight	Weight in grams	
1	2	8.2	98	243	SD-2
2	4	9.3	96	218.3	SD-4
3	6	10.5	94	198.5	SD-6

Materials Joining and Manufacturing Processes: MJMP 2025   Materials Research Forum LLC
Materials Research Proceedings 55 (2025) 156-161   https://doi.org/10.21741/9781644903612-23

Results and Discussion

Tensile test results

Figure 1(a) presents the stress-strain curves of PBAT-sawdust composites, revealing their brittle nature with no yield points before fracture.

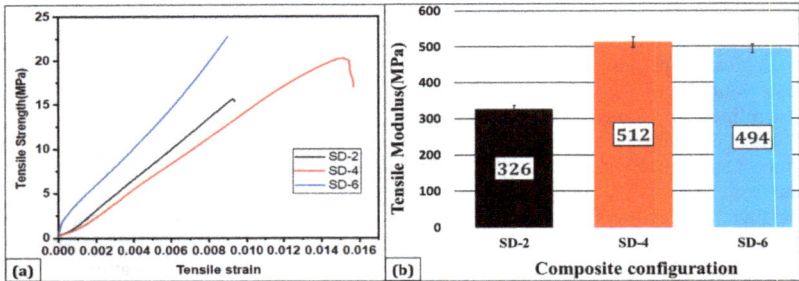

Fig.1: (a) Tensile strength versus tensile strain curves of various SD-PBAT composites and (b) Tensile modulus versus composite configuration

Among them, SD-4 exhibits the highest tensile stress, surpassing SD-2 by 30% and SD-6 by 9%. Figure 8(b) compares the tensile modulus across composites, showing that SD-4 has a significantly higher modulus—36% greater than SD-2 and 4% higher than SD-6—indicating improved stiffness. In the SD-4 composite, the increased tensile strength is attributed to the uniform dispersion of sawdust particles within the PBAT matrix [11], resulting in strong interfacial adhesion between sawdust and PBAT [12]. However, as the sawdust content increases, tensile strength decreases due to particle agglomeration, which weakens interfacial adhesion [13], leading to reduced tensile strength in the SD-6 sample.

Flexural test results

Figure 2(a) presents load-deformation curves for different composite configurations, showing that SD-4 has the highest load-carrying capacity, surpassing SD-2 by 23% and SD-6 by 11%. Figure 2(b) illustrates flexural modulus versus composite configuration, where SD-4 exhibits an 18% and 13% higher modulus than SD-2 and SD-6.

Materials Joining and Manufacturing Processes: MJMP 2025
Materials Research Proceedings 55 (2025) 156-161

Materials Research Forum LLC
https://doi.org/10.21741/9781644903612-23

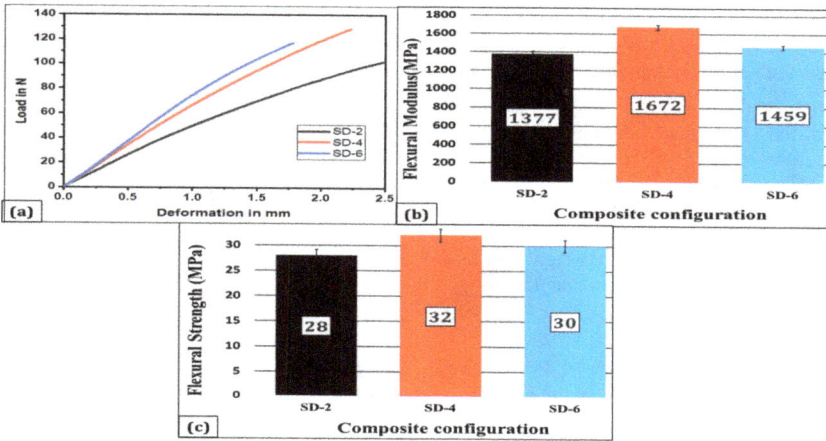

Fig.2: (a) Load versus deformation curve (b) Flexural modulus versus composite configuration and (c) Flexural strength versus composite configuration

Figure 2(c) presents flexural strength versus composite configuration, indicating SD-4 is approximately 13% stronger than SD-2 and 6% stronger than SD-6 respectively, demonstrating improved stiffness and load resistance. As the saw dust particle increases in the matrix the stiffness of the composite increases and reaches maximum strength and modulus at 4 wt.% which attributed to the increased stiffness of the filler content in the composite. Further increase in filler content decreases the flexural strength due poor dispersion of filler particles into the matrix [14].

Impact test results
Figure 3(b) shows absorbed energy versus composite configuration which indicates that SD-4 composite has capacity to absorb energy upon impact of about 39%, and 4% greater than that of SD-2, and SD-6 composites. Figure 3(a) shows impact strength versus composite configuration indicating SD-4 has a greater impact strength with values of about 27%, and 12% than composites constructed with SD-2, and SD-6 composition, respectively.

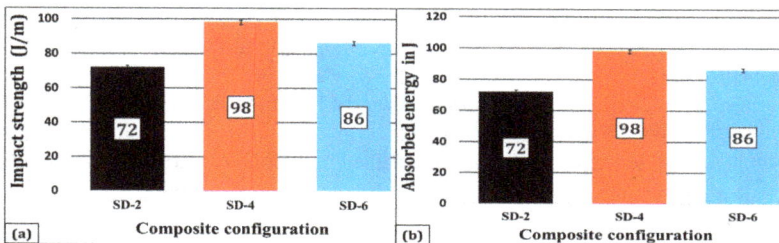

Fig.3: (a) Impact strength versus composite configuration and (b) absorbed energy versus composite configuration

Materials Joining and Manufacturing Processes: MJMP 2025   Materials Research Forum LLC
Materials Research Proceedings 55 (2025) 156-161   https://doi.org/10.21741/9781644903612-23

Impact strength of SD-4 composite is highest due to uniform dispersion of filler particles, moreover the impact strength improvement is due to PBAT matrix which contains soft elastomeric phase with excellent toughness and flexibility [15, 16].

Conclusion
This study investigated the mechanical performance of 3D-printed PBAT composites reinforced with sawdust at varying concentrations (2%, 4%, and 6%). The results indicate that incorporating 4% sawdust provides optimal mechanical properties, achieving the highest tensile, flexural, and impact strengths. Increased sawdust content beyond 4% leads to particle agglomeration, reducing interfacial adhesion and mechanical performance. These findings highlight the potential of PBAT-sawdust composites for lightweight and biodegradable applications, offering a balance between strength, flexibility, and sustainability.

References

[1] Maruthi Prashanth, B. H., M. Prashanth Pai, C. V. Pujar, I. M. "Understanding the Influence of Amino Resin Percentage on the Hybrid Abaca-Kenaf Polyester Composite's Mechanical Properties." In International Conference on Smart and Sustainable Developments in Materials, Manufacturing and Energy Engineering, pp. 29-34. Cham: Springer Nature Switzerland, 2023. https://doi.org/10.1007/978-3-031-63909-8_5

[2] , S.; Pereira, S.; Juliana, M.F.S.; Bluma, G.S.; Fully biodegradable composites based on poly (butylene adipate-coterephthalate)/peach palm trees fiber. Compos. Part B 2017, 129, 117-123. https://doi.org/10.1016/j.compositesb.2017.07.088

[3] Arvind, G.; Bansri, C.; Boon, P.C.; Tizazu, M. Robust and sustainable PBAT-Hemp residue biocomposites: Reactive extrusion compatibilization and fabrication. Compos. Sci. Technol. 2021, 215, 109014. https://doi.org/10.1016/j.compscitech.2021.109014

[4] Musioł, Marta, Joanna Rydz, Henryk Janeczek, 2021. "(Bio)Degradable Biochar Composites - Studies on Degradation and Electrostatic Properties." Materials Science and Engineering B 275 (November): 115515-15. https://doi.org/10.1016/j.mseb.2021.115515

[5] Yang, Xuping, and Shengyuan Zhong. 2020. "Properties of Maleic Anhydride-Modified Lignin Nanoparticles/Polybutylene Adipate-Co-Terephthalate Composites." Journal of Applied Polymer Science 137 (35): 49025. https://doi.org/10.1002/app.49025

[6] Vinicius Carrillo Beber, Silvio de Barros, M D Banea, M Brede, Laura H Carvalho, Ron Hoffmann, Anna Raffaela, et al. 2018. "Effect of Babassu Natural Filler on PBAT/PHB Biodegradable Blends: An Investigation of Thermal, Mechanical, and Morphological Behavior" 11 (5): 820-20. https://doi.org/10.3390/ma11050820

[7] Wataya, Célio H, Roberta A Lima, Rene R Oliveira, and B Moura. 2015. "Mechanical, Morphological and Thermal Properties of Açaí Fibers Reinforced Biodegradable Polymer Composites." Characterization of Minerals, Metals, and Materials, January, 265-72. https://doi.org/10.1007/978-3-319-48191-3_33

[8] ASTM D3039/D3039M-14, Standard Test Method for Tensile Properties of Polymer Matrix Composite Materials. Vol. D15.03.

[9] ASTM D790-03, Standard Test Methods for Flexural Properties of Unreinforced and Reinforced Plastics and Electrical Insulating Materials. Vol. D08.03.

Materials Joining and Manufacturing Processes: MJMP 2025
Materials Research Proceedings 55 (2025) 156-161

Materials Research Forum LLC
https://doi.org/10.21741/9781644903612-23

[10] ASTM D5045-14, Standard Test Methods for Plane-Strain Fracture Toughness and Strain Energy Release Rate of Plastic Materials. "Test Methods for Plane-Strain Fracture Toughness and Strain Energy Release Rate of Plastic Materials."

[11] A.A. Okhlopkova, L.A Nikiforov, and R.V Borisova. 2016. "Polymer Nanocomposites Exploited under the Arctic Conditions." KnE Materials Science 1 (1): 122-22. https://doi.org/10.18502/kms.v1i1.573

[12] Ming-Zhu, Pan, Mei Chang-Tong, Zhou Xu-Bing, and Pu Yun-Lei. 2011. "Effects of Rice Straw Fiber Morphology and Content on the Mechanical and Thermal Properties of Rice Straw Fiber-High Density Polyethylene Composites." Journal of Applied Polymer Science 121 (5): 2900-2907. https://doi.org/10.1002/app.33913

[13] , Koay Seong, Salmah Husseinsyah, and Hakimah Osman. 2012. "Mechanical and Thermal Properties of Coconut Shell Powder Filled Polylactic Acid Biocomposites: Effects of the Filler Content and Silane Coupling Agent." Journal of Polymer Research 19 (5). https://doi.org/10.1007/s10965-012-9859-8

[14] Jurczyk, Sebastian, Jacek Andrzejewski, Adam Piasecki, Marta Musioł, Joanna Rydz, and Marek Kowalczuk. 2024. "Mechanical and Rheological Evaluation of Polyester-Based Composites Containing Biochar." Polymers 16 (9): 1231-31. https://doi.org/10.3390/polym16091231

[15] Goriparthi, Bhanu K., K.N.S. Suman, and Mohan Rao Nalluri. 2011. "Processing and Characterization of Jute Fiber Reinforced Hybrid Biocomposites Based on Polylactide/Polycaprolactone Blends." Polymer Composites 33 (2): 237-44. https://doi.org/10.1002/pc.22145

[16] CG, Ramachandra, and Prashanth pai. "The Effect of Metal Filler on the Mechanical Performance of Epoxy Resin Composites." Engineering Proceedings 59, no. 1 (2024): 200. https://doi.org/10.3390/engproc2023059200

Keyword Index

3D Printing 19, 72, 118, 156

AA6082 34
AA7075, Friction Stir Welding 95
AA8009 102
ABS 19
Additive Manufacturing 72
Al-15Sn 7
Alloying Element 7
Alloying 136
Aluminium 130
ANSYS 110
AUTODYN 27
Automation in Dye Penetrant Test 40

Biocompatible 27
Buckling and FE Analysis 45

Carbide Tools 12
Coefficient of Thermal Expansion 64
Cold and Warm Forging 34
Composite Laminates 45
Copper (Cu) Nanoparticles 57
Corrosion 141
Cubic Infill 118

Difficult-to-Cut Material 1
Discontinuity 40
Dye/Florescent Techniques 40

Electrically Assisted Forming (EAF) 85
Electromagnetic 130
Embedded Cutouts 45
Epoxy Resin (ER) 57
Epoxy Resin Composites 64
Explosive Welding 51

FDM 118
Filament 19
Filler Particles 64
Fired Boilers 141
Forging Equipment 34
Friction Stir Welding 102

Glass Fibre (GF) 57, 64
Graphene Nanoplatelets 95

Hardness 102
Heat Treatment 7
High Chrome White Cast Iron (HCWCI) 12
High Temperature 136
Honeycomb Infill 118

Implosive Reactive Armour 79
Industry 4.0 40
Industry Applications 72

Joggle Joining 79
Joining 130

Laser Shock Forming 85

Machine Learning 51
Magnesium (Mg) 27
Manganese (Mn) Nanoparticles 57
Materials Science 72
Mechanical Properties 95, 156
Mechanical 7
Micro Forming 85
Micro Hardness 148
Micro Structure 148
Micro-Manufacturing 85
Microstructure 7, 95, 102
Microwave Welding 110

Milling	1
Modern Warfare	79
Monel-400	110
Natural Fiber	19
Optimization	51
Pb Free Bearing Material	7
Pb-Free Solders	136
PLA	19, 118
Polybutylene Adipate Terephthalate	156
Pressure	27
PVC	130
Quenching	7
Residual Stress	130
Sawdust	156
Silicon Carbide (SiC) Nanoparticles	57
Smoothed Particle Hydrodynamics (SPH)	27
Statistical Analysis	72
Strain	27
Structural Analysis	110
Structure-Property Correlation	34
Surface Roughness	1
Taguchi Optimization	118
Techniques	51
Tensile Strength	102
Tensile	45
Tension	19
Thermal Analysis	110
Thermal Conductivity	57, 64
Thermal Contact Resistance (TCR)	148
Thermal Interface Material (TIM)	148

Thermal Spray Coating	141
Thermogravimetric Analysis (TGA)	57, 64
Tool Wear	1
Ultrasonic Vibration Assisted (UVA) Micro Forming	85
Velocity	27
Wavy Interface	51
Wear Properties	7
Wetting	136
Wire Arc Additive Manufacturing	1
Zinc (Zn)	27

About the Editors

Dr. Satyanarayan is currently working as Professor and Head, Dept. of Mechanical Engineering, Alva's Institute of Engineering Technology (AIET), Moodubidire, India. He has been working at AIET since July 2013. He did his B. E (2004) in Mechanical Engineering from Basaveshwar Engg College, Bagalkot, M. Tech (2009) and Ph.D (2014) from NIT Surathkal, Karnataka. He was a Post–Doctoral Researcher at IIPS (Dec 2015 –Jan 2016), and IINa (Nov 2023 – March 2024) at Kumamoto University, Japan. He has published more than 50 research papers in Scopus-indexed journals and presented more than 30 research papers in International conferences. Two granted Indian patents credited him. Dr. Satyanarayan written three book chapters, reviewed more than 300 research papers and received research grants from CPRI, Govt. of India, VGST and KSCST, Govt. of Karnataka. He was a guest editor (two times) for materials today: proceedings, Elsevier publications, editorial board member for premier journals, and technical advisory committee member for several international conferences. He has organised three international conferences and seven national workshops. More than 15 invited technical talks were delivered by him. He guided Ph.D and M.Tech and UG thesis. Received best paper awards in international conferences (04) and best project award in the exhibition by KSCST, Govt. of Karnataka.

Dr. Kazuyuki Hokamoto is currently working as a Specially Appointed Professor at Institute of Industrial Nanomaterials (IINa), Kumamoto University, Japan after his retirement from IINa in March, 2025. He began his faculty position at the Department of Mechanical Engineering, Kumamoto University, Japan, in 1987 followed by being awarded his doctoral degree (Dr. Eng.) from Kyushu University in 1988. He has served as a faculty member of Kumamoto University, and he served as the Deputy–Director of the Institute of Pulsed Power Science (IPPS), Kumamoto University. IPPS was reorganized to IINa in 2020 and he has been served as the Head of Explosion & Impulsive Processing Laboratory at IINa until his retirement. His research interest is mainly on the use of explosive and other impulsive energy for materials processing, and he has published about 200 papers mainly on explosive welding, powder compaction and synthesis of new materials. His experiences include positions as Visiting Scholar at the University of California, San Diego (1989–90), and Georgia Institute of Technology (1995–96). He has organized the ESHP Symposium series several times since 2004. He has been awarded Best Paper Award in 2004 from Japan Soc. for Technol. of Plasticity, Promotion of Science Award in 2016 from Japan Welding Society, Glass Memorial Lecture Award in 2025 from Shock Wave Society, Japan, and others. He guided several Ph.D, Masters and UG thesises including foreign students.

Dr. Suresh P. S. is a seasoned academician and researcher with over 21 years of combined experience in academia and industry. He holds a Ph.D. in Mechanical Engineering, with a specialization in natural fiber-reinforced polymer composites. He completed his M.Tech in Industrial Automation and Robotics, and B.E. in Industrial and Production Engineering from Malnad College of Engineering, Hassan. Dr. Suresh currently serves as a Senior Assistant Professor in the Department of Mechanical Engineering at Alva's Institute of Engineering & Technology, Moodbidri. He plays a pivotal role in academic coordination and institutional development and is actively involved in organizing national and international conferences. Dr. Suresh has published research articles in reputed journals, reviewed manuscripts for international journals, and presented his research at various international conferences. He is proficient in industry-standard software tools and simulation platforms. Dr. Suresh is a lifetime member of the Indian Society for Technical Education (ISTE) and the Indian Red Cross Society, underscoring his professional affiliation with both academic and service communities.

Dr. Kumar Swamy M.C. is a committed academic professional with extensive experience in both teaching and research. He has been serving as a Senior Assistant Professor at Alva's Institute of Engineering and Technology, Moodbidri. As a Life Member of the Indian Society of Technical Education (L.M.I.S.T.E), Dr. Kumar plays a significant role in shaping the academic environment at his institution. He specializes in teaching core mechanical engineering subjects such as Basic Thermodynamics, Applied Thermodynamics, and Heat and Mass Transfer, providing students with a strong foundation in these crucial areas. In addition to his teaching responsibilities, Dr. Kumar is actively involved in fostering industry-academia collaboration, organizing campus placement activities, and overseeing student training programs to ensure they are well-prepared for the workforce. His research interests span across IC engines, gasification, computational fluid dynamics (CFD), interfacial materials, and electronic cooling systems. Dr. Kumar has presented his research at several prestigious international conferences, and his work has been published in well-regarded journals, further contributing to the body of knowledge in his field.

e

www.ingramcontent.com/pod-product-compliance
Lightning Source LLC
Chambersburg PA
CBHW071234210326
41597CB00016B/2048

* 9 7 8 1 6 4 4 9 0 3 6 0 5 *